ELECTRICITY AND MAGNETISM

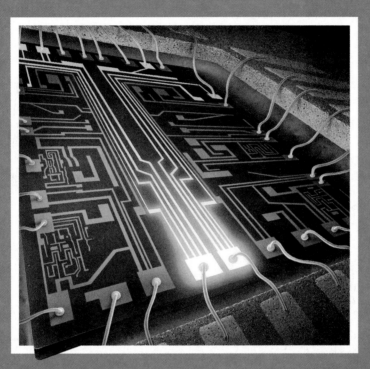

Anthea Maton
Former NSTA National Coordinator
Project Scope, Sequence, Coordination
Washington, DC

Jean Hopkins
Science Instructor and Department Chairperson
John H. Wood Middle School
San Antonio, Texas

Susan Johnson
Professor of Biology
Ball State University
Muncie, Indiana

David LaHart
Senior Instructor
Florida Solar Energy Center
Cape Canaveral, Florida

Charles William McLaughlin
Science Instructor and Department Chairperson
Central High School
St. Joseph, Missouri

Maryanna Quon Warner
Science Instructor
Del Dios Middle School
Escondido, California

Jill D. Wright
Professor of Science Education
Director of International Field Programs
University of Pittsburgh
Pittsburgh, Pennsylvania

 Prentice Hall
Englewood Cliffs, New Jersey
Needham, Massachusetts

Prentice Hall Science
Electricity and Magnetism

Student Text and Annotated Teacher's Edition
Laboratory Manual
Teacher's Resource Package
Teacher's Desk Reference
Computer Test Bank
Teaching Transparencies
Science Reader
Product Testing Activities
Computer Courseware
Video and Interactive Video

The illustration on the cover, rendered by Keith Kasnot, shows an integrated circuit board used in computers and many other electronic devices.

Credits begin on page 129.

FIRST EDITION

ISBN 0-13-981044-7

3 4 5 6 7 8 9 10   96 95 94 93

Prentice Hall
A Division of Simon & Schuster
Englewood Cliffs, New Jersey 07632

STAFF CREDITS

Editorial:	Harry Bakalian, Pamela E. Hirschfeld, Maureen Grassi, Robert P. Letendre, Elisa Mui Eiger, Lorraine Smith-Phelan, Christine A. Caputo
Design:	AnnMarie Roselli, Carmela Pereira, Susan Walrath, Leslie Osher, Art Soares
Production:	Suse Cioffi, Joan McCulley, Elizabeth Torjussen, Christina Burghard, Marlys Lehmann
Photo Research:	Libby Forsyth, Emily Rose, Martha Conway
Publishing Technology:	Andrew Grey Bommarito, Gwendollynn Waldron, Deborah Jones, Monduane Harris, Michael Colucci, Gregory Myers, Cleasta Wilburn
Marketing:	Andy Socha, Victoria Willows
Pre-Press Production:	Laura Sanderson, Denise Herckenrath
Manufacturing:	Rhett Conklin, Gertrude Szyferblatt

Consultants

Kathy French	National Science Consultant
William Royalty	National Science Consultant

Contributing Writers

Linda Densman
Science Instructor
Hurst, TX

Linda Grant
Former Science Instructor
Weatherford, TX

Heather Hirschfeld
Science Writer
Durham, NC

Marcia Mungenast
Science Writer
Upper Montclair, NJ

Michael Ross
Science Writer
New York City, NY

Content Reviewers

Dan Anthony
Science Mentor
Rialto, CA

John Barrow
Science Instructor
Pomona, CA

Leslie Bettencourt
Science Instructor
Harrisville, RI

Carol Bishop
Science Instructor
Palm Desert, CA

Dan Bohan
Science Instructor
Palm Desert, CA

Steve M. Carlson
Science Instructor
Milwaukie, OR

Larry Flammer
Science Instructor
San Jose, CA

Steve Ferguson
Science Instructor
Lee's Summit, MO

Robin Lee Harris
Freedman
Science Instructor
Fort Bragg, CA

Edith H. Gladden
Former Science Instructor
Philadelphia, PA

Vernita Marie Graves
Science Instructor
Tenafly, NJ

Jack Grube
Science Instructor
San Jose, CA

Emiel Hamberlin
Science Instructor
Chicago, IL

Dwight Kertzman
Science Instructor
Tulsa, OK

Judy Kirschbaum
Science/Computer Instructor
Tenafly, NJ

Kenneth L. Krause
Science Instructor
Milwaukie, OR

Ernest W. Kuehl, Jr.
Science Instructor
Bayside, NY

Mary Grace Lopez
Science Instructor
Corpus Christi, TX

Warren Maggard
Science Instructor
PeWee Valley, KY

Della M. McCaughan
Science Instructor
Biloxi, MS

Stanley J. Mulak
Former Science Instructor
Jensen Beach, FL

Richard Myers
Science Instructor
Portland, OR

Carol Nathanson
Science Mentor
Riverside, CA

Sylvia Neivert
Former Science Instructor
San Diego, CA

Jarvis VNC Pahl
Science Instructor
Rialto, CA

Arlene Sackman
Science Instructor
Tulare, CA

Christine Schumacher
Science Instructor
Pikesville, MD

Suzanne Steinke
Science Instructor
Towson, MD

Len Svinth
Science Instructor/
Chairperson
Petaluma, CA

Elaine M. Tadros
Science Instructor
Palm Desert, CA

Joyce K. Walsh
Science Instructor
Midlothian, VA

Steve Weinberg
Science Instructor
West Hartford, CT

Charlene West, PhD
Director of Curriculum
Rialto, CA

John Westwater
Science Instructor
Medford, MA

Glenna Wilkoff
Science Instructor
Chesterfield, OH

Edee Norman Wiziecki
Science Instructor
Urbana, IL

Teacher Advisory Panel

Beverly Brown
Science Instructor
Livonia, MI

James Burg
Science Instructor
Cincinnati, OH

Karen M. Cannon
Science Instructor
San Diego, CA

John Eby
Science Instructor
Richmond, CA

Elsie M. Jones
Science Instructor
Marietta, GA

Michael Pierre
McKereghan
Science Instructor
Denver, CO

Donald C. Pace, Sr.
Science Instructor
Reisterstown, MD

Carlos Francisco Sainz
Science Instructor
National City, CA

William Reed
Science Instructor
Indianapolis, IN

Multicultural Consultant

Steven J. Rakow
Associate Professor
University of Houston—
Clear Lake
Houston, TX

English as a Second Language (ESL) Consultants

Jaime Morales
Bilingual Coordinator
Huntington Park, CA

Pat Hollis Smith
Former ESL Instructor
Beaumont, TX

Reading Consultant

Larry Swinburne
Director
Swinburne Readability
Laboratory

CONTENTS

ELECTRICITY AND MAGNETISM

Reference Section

Features

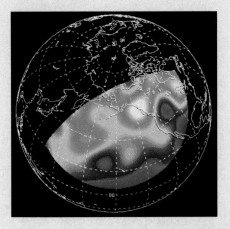

CONCEPT MAPPING

Throughout your study of science, you will learn a variety of terms, facts, figures, and concepts. Each new topic you encounter will provide its own collection of words and ideas—which, at times, you may think seem endless. But each of the ideas within a particular topic is related in some way to the others. No concept in science is isolated. Thus it will help you to understand the topic if you see the whole picture; that is, the interconnectedness of all the individual terms and ideas. This is a much more effective and satisfying way of learning than memorizing separate facts.

Actually, this should be a rather familiar process for you. Although you may not think about it in this way, you analyze many of the elements in your daily life by looking for relationships or connections. For example, when you look at a collection of flowers, you may divide them into groups: roses, carnations, and daisies. You may then associate colors with these flowers: red, pink, and white. The general topic is flowers. The subtopic is types of flowers. And the colors are specific terms that describe flowers. A topic makes more sense and is more easily understood if you understand how it is broken down into individual ideas and how these ideas are related to one another and to the entire topic.

It is often helpful to organize information visually so that you can see how it all fits together. One technique for describing related ideas is called a **concept map**. In a concept map, an idea is represented by a word or phrase enclosed in a box. There are several ideas in any concept map. A connection between two ideas is made with a line. A word or two that describes the connection is written on or near the line. The general topic is located at the top of the map. That topic is then broken down into subtopics, or more specific ideas, by branching lines. The most specific topics are located at the bottom of the map.

To construct a concept map, first identify the important ideas or key terms in the chapter or section. Do not try to include too much information. Use your judgment as to what is really important. Write the general topic at the top of your map. Let's use an example to help illustrate this process. Suppose you decide that the key terms in a section you are reading are School, Living Things, Language Arts, Subtraction, Grammar, Mathematics, Experiments, Papers, Science, Addition, Novels. The general topic is School. Write and enclose this word in a box at the top of your map.

SCHOOL

Now choose the subtopics—Language Arts, Science, Mathematics. Figure out how they are related to the topic. Add these words to your map. Continue this procedure until you have included all the important ideas and terms. Then use lines to make the appropriate connections between ideas and terms. Don't forget to write a word or two on or near the connecting line to describe the nature of the connection.

Do not be concerned if you have to redraw your map (perhaps several times!) before you show all the important connections clearly. If, for example, you write papers for Science as well as for Language Arts, you may want to place these two subjects next to each other so that the lines do not overlap.

One more thing you should know about concept mapping: Concepts can be correctly mapped in many different ways. In fact, it is unlikely that any two people will draw identical concept maps for a complex topic. Thus there is no one correct concept map for any topic! Even

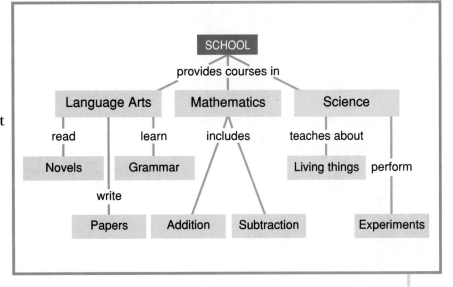

though your concept map may not match those of your classmates, it will be correct as long as it shows the most important concepts and the clear relationships among them. Your concept map will also be correct if it has meaning to you and if it helps you understand the material you are reading. A concept map should be so clear that if some of the terms are erased, the missing terms could easily be filled in by following the logic of the concept map.

ELECTRICITY AND MAGNETISM

▲ As if made of flesh and blood, these computer-controlled dinosaurs appear to be munching on assorted fruits.

Huge dinosaurs roar wildly as they tower over your head. Some cause the ground to tremble merely by walking past you. Have you been transported back in time? No. You are at Epcot Center in Walt Disney World where animated creatures are designed and controlled by computers.

Although this computer application is specifically designed to amuse and entertain you, other computer applications have more scientific and informational uses. Researchers can enter available data into computers in order to build complete, moving models of subjects that can only be imagined—not observed. For example, computers have enabled scientists to design a wide range of new products, simulate various aspects of prehistoric life, and hypothesize about the features of distant planets and galaxies. The beauty of such applications is that designers—whether they are interested in the past, present, or future—can build, experiment with, and improve upon models with the speed, accuracy, and safety of a computer and its display devices.

◀ The design of many new products is aided by computers. In this way designers can observe a new product, such as this shoe, and test various features of the product before it is manufactured.

CHAPTERS

Much of the technology that makes computer applications possible is more familiar than you might think. In this textbook you will discover that the phenomona of electricity and magnetism reach far beyond household appliances and industrial devices. They reach into the heart of present and future technology.

It would be quite costly for automobile manufacturers to constantly build test models of new cars. Instead, computers enable designers to create models on a screen. This computer model is being tested for aerodynamic efficiency. ▶

Discovery *Activity*

Flying Through Air With the Greatest of Ease

1. Inflate a balloon and tie the end.

2. Bring the balloon within centimeters of a plate filled with a mixture of salt and pepper. Observe for 3 to 5 minutes.

3. Take the balloon away from the plate. Rub the balloon with a piece of wool. Repeat step 2.

   ■ How is the behavior of the salt and pepper different before and after you rub the balloon with wool?

4. Arrange small piles of each of the following materials: paper clips, rubber bands, pieces of paper, assorted metal and plastic buttons. Make sure that each of the piles is separated from the others.

5. Hold a magnet above the first pile without touching it. Observe what happens. Repeat this procedure over each of the remaining piles.

   ■ What happens to each of the materials when you place the magnet above it? What can you conclude about how different materials are affected by a magnet?

Electric Charges and Currents

Creepy characters . . . dark nights . . . thunder and lightning crashing in the background . . . castles with trap doors and secret laboratories. Do these descriptions sound familiar to you? Perhaps you have seen them in monster movies such as *Frankenstein* and *The Bride of Frankenstein*. These exciting movies often express people's hidden hopes and fears about a world in which scientific knowledge can be used for either good or evil. Usually, electricity is used at some point in the movie to mysteriously create life or to destroy it.

For hundreds of years, many people were frightened by electricity and believed it to have mysterious powers. Today a great deal is known about electricity. And although it is not mysterious, electricity plays a powerful role in your world. Electricity is involved in all interactions of everyday matter—from the motion of a car to the movement of a muscle to the growth of a tree. Electricity makes life easier and more comfortable. In this chapter you will discover what electricity is, how it is produced and used, and why it is so important.

Journal *Activity*

You and Your World Did you switch on a light, shut off an alarm clock, listen to the radio, or turn on a hair dryer today? In your journal, describe the importance of electricity in your daily life. Include any questions you may have about electricity.

◀ *Dr. Frankenstein at work in his laboratory.*

1–1 Electric Charge

Have you ever rubbed a balloon on your clothing to make it stick to you or to a wall? Or have you ever pulled your favorite shirt out of the clothes dryer only to find socks sticking to it? How can objects be made to stick to one another without glue or tape? Believe it or not, the answer has to do with electricity. And the origin of electricity is in the particles that make up matter.

Atoms and Electricity

All matter is made of **atoms.** An atom is the smallest particle of an element that has all the properties of that element. An element contains only one kind of atom. For example, the element lead is made of only lead atoms. The element gold is made of only gold atoms.

Atoms themselves are made of even smaller particles. Three of the most important particles are **protons, neutrons,** and **electrons.** Protons and neutrons are found in the nucleus, or center, of an atom. Electrons are found in an area outside the nucleus often described as an electron cloud. **Both protons and electrons have a basic property called charge.** Unlike many other physical properties of matter, **charge** is not something you can see, weigh, or define. However, you can observe the effects of charge—more specifically, how charge affects the behavior of particles.

The magnitude, or size, of the charge on the proton is the same as the magnitude of the charge on the electron. The kind of charge, however, is not the same for both particles. Protons have a positive charge, which is indicated by a plus symbol (+). Electrons have a negative charge, which is indicated by a minus symbol (−). Neutrons are neutral, which means that they have no electric charge. The terms positive and negative, which have no real physical

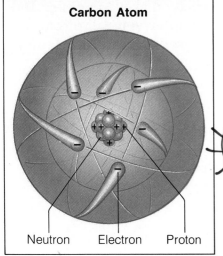

Carbon Atom

Neutron   Electron   Proton

Figure 1–1 *The diagram of the carbon atom shows the arrangement of subatomic particles known as protons, neutrons, and electrons. Carbon is found in all living organisms, including this hungry hippo.*

significance, were originally decided upon by Benjamin Franklin when he first discovered charge. They have been used ever since.

Charge and Force

The difference between the two charges has to do with how they behave and the **forces** they exert. A force is a pull or push on an object. You are already familiar with various types of forces. Your foot exerts a force (a push) on a ball when you kick it. An ocean wave exerts a force (a push) on you when it knocks you over. The Earth exerts a force (a pull) on the moon to keep it in its orbit. Charged particles exert similar pushes and pulls.

When charged particles come near one another, they give rise to two different forces. A force that pulls objects together is a force of attraction. **A force of attraction exists between oppositely charged particles.** So negatively charged electrons are attracted to positively charged protons. This force of attraction holds the electrons in the electron cloud surrounding the nucleus.

A force that pushes objects apart is a force of repulsion. **A force of repulsion exists between particles of the same charge.** Negatively charged electrons repel one another, just as positively charged protons do. Electric charges behave according to this simple rule: *Like charges repel each other; unlike charges attract each other.*

FIND OUT BY DOING

Electric Forces

1. Take a hard rubber (not plastic) comb and rub it with a woolen cloth.

2. Bring the comb near a small piece of cork that is hanging from a support by a thread.

3. Allow the comb to touch the cork, and then take the comb away. Bring the comb toward the cork again.

4. Repeat steps 1 to 3 using a glass rod rubbed with silk. Then bring the rubber comb rubbed with wool near the cork.

■ Record and explain your observations.

Figure 1–2 *When charged particles come near each other, a force is produced. The force can be either a force of attraction or a force of repulsion. What is the rule of electric charges?*

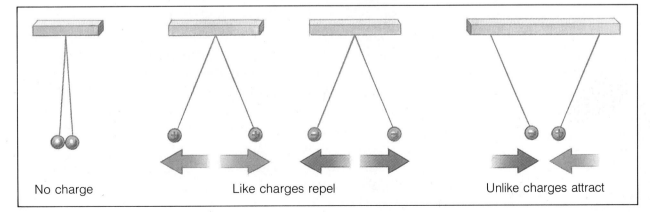

No charge   Like charges repel   Unlike charges attract

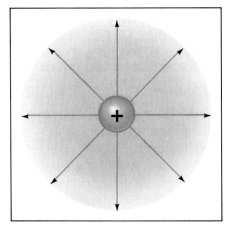

Figure 1–3 *Lines of force show the nature of the electric field surrounding a charged particle. When two charged particles come near each other, the electric fields of both particles are altered as shown.*

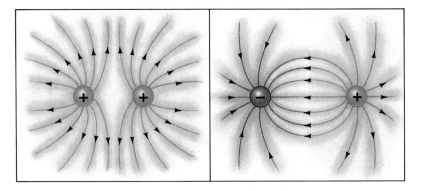

Electric Field

The attraction and repulsion of charged particles occurs because charged particles have electric fields around them. An **electric field** is an area over which an electric charge exerts a force. When a charged particle moves into the electric field of another charged particle, it is either pushed or pulled depending on the relationship between the two particles.

The electric field is the strongest near the charged particle. As the distance from the charged particle increases, the strength of the electric field decreases. As shown in Figure 1–3, the electric field can be visualized by drawing lines extending outward from a charged particle.

1–1 Section Review

1. Describe the charged particles in an atom.
2. What is a force? Give some examples.
3. What is the rule of electric charges?
4. What is an electric field?

Critical Thinking—*Drawing Diagrams*
5. A positively charged particle is placed 1 centimeter from positively charged particle X. Describe the forces experienced by each particle. Compare these forces with the forces that would exist if a negatively charged particle were placed 10 centimeters from particle X. Draw the electric field surrounding each particle.

1–2 Static Electricity

From your experience, you know that when you sit on a chair, pick up a pen, or put on your jacket, you are neither attracted nor repelled by these objects. Although the protons and electrons in the atoms of these objects have electric charges, the objects themselves are neutral. Why?

An atom has an equal number of protons and electrons. So the total positive charge is equal to the total negative charge. The charges cancel out. So even though an atom contains charged particles, it is electrically neutral. It has no overall charge.

How then do objects such as balloons and clothing develop an electric charge if these objects are made of neutral atoms? The answer lies in the fact that electrons, unlike protons, are free to move. In certain materials, some of the electrons are only loosely held in their atoms. Thus these electrons can easily be separated from their atoms. If an atom loses an electron, it becomes positively charged because it is left with more positive charges (protons) than negative charges (electrons). If an atom gains an electron, it becomes negatively charged. Why? An atom that gains or loses electrons is called an ion.

Guide for Reading

Focus on these questions as you read.

▶ How do neutral objects acquire charge?

▶ What is static electricity?

Figure 1–4 *Is it magic that makes these pieces of paper rise up to the comb? No, just static electricity. Have you ever experienced static electricity?*

FIND OUT BY

DOING

Balloon Electricity

1. Blow up three or four medium-sized balloons.

2. Rub each balloon vigorously on a piece of cloth. Wool works especially well.

3. "Stick" each balloon on the wall. Record the day, time, and weather conditions.

4. Every few hours, check the position of the balloons.

5. Repeat your experiment on a day when the weather conditions are different—for example, on a dry day versus on a humid or rainy day.

How long did the balloons stay attached to the wall? Why did they eventually fall off the wall?

■ Does weather have any effect? Explain.

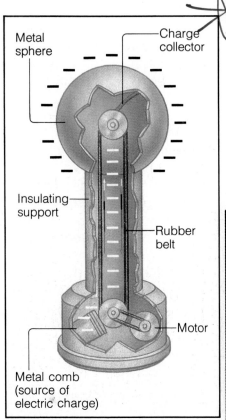

Metal sphere

Charge collector

Insulating support

Rubber belt

Motor

Metal comb (source of electric charge)

Just as an atom can become a negatively or positively charged ion, so can an entire object acquire a charge. **A neutral object acquires an electric charge when it either gains or loses electrons.** Remember, only electrons move. Also remember that charge is neither being created nor destroyed. Charge is only being transferred from one object to another. This is known as the Law of Conservation of Charge.

Methods of Charging

When you rub a balloon against a piece of cloth, the cloth loses some electrons and the balloon gains these electrons. The balloon is no longer a neutral object. It is a negatively charged object because it has more electrons than protons. As the negatively charged balloon approaches a wall, it repels the electrons in the wall. The electrons in the area of the wall nearest the balloon move away, leaving that area of the wall positively charged. See Figure 1–6. Using the rule of charges, can you explain why the balloon now sticks to the wall?

Rubbing two objects together is one method by which an object can become charged. This method is known as the **friction** method. In the previous example, the balloon acquired a charge by the friction method. That is, it was rubbed against cloth.

Figure 1–5 *A Van de Graaff generator produces static electricity by friction. Electrons ride up a rubber belt to the top of the generator, where they are picked off and transferred to the metal sphere. The charge that has built up on the generator at the Ontario Science Center is large enough to make this girl's hair stand on end.*

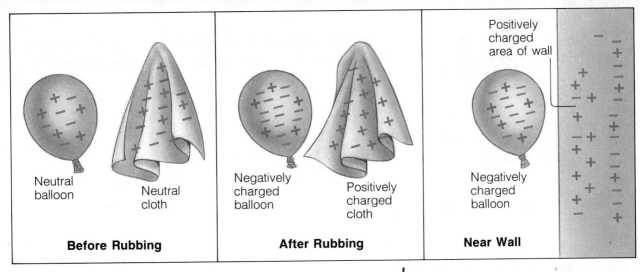

		Positively charged area of wall
Neutral balloon	Neutral cloth	
	Negatively charged balloon	Positively charged cloth
		Negatively charged balloon
Before Rubbing	**After Rubbing**	**Near Wall**

Another method of charging is **conduction.** In conduction, which involves the direct contact of objects, electrons flow through one object to another object. Certain materials permit electric charges to flow freely. Such materials are called **conductors.** Most metals are good conductors of electricity. This is because some electrons in the atoms are free to move throughout the metal. Silver, copper, aluminum, and mercury are among the best conductors. The Earth is also a good conductor.

Materials that do not allow electric charges to flow freely are called **insulators.** Insulators do not conduct electric charges well because the electrons in the atoms of insulators are tightly bound and cannot move throughout the material. Good insulators include rubber, glass, wood, plastic, and air. The rubber tubing around an electric wire and the plastic handle on an electric power tool are examples of insulators.

Figure 1–6 *Rubbing separates charges, giving the cloth a positive charge and the balloon a negative charge. When the negatively charged balloon is brought near the wall, it repels electrons in the wall. The nearby portion of the wall becomes positively charged. What happens next?*

Figure 1–7 *A metal rod can be charged negatively (left) or positively (right) by conduction.*

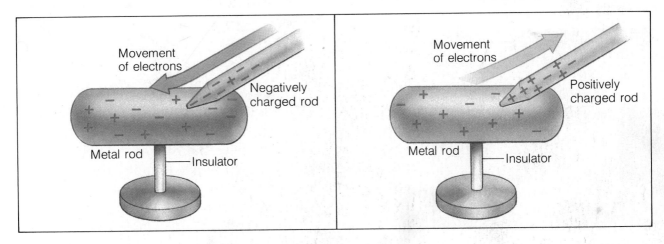

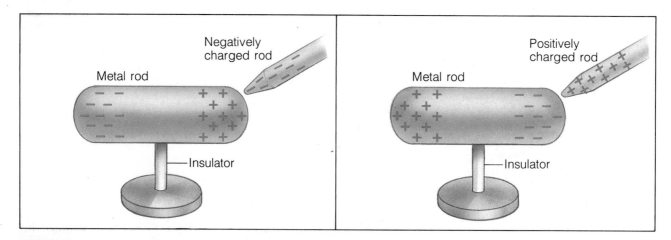

Figure 1-8 *A charged rod brought near a conductor induces an electric charge in the conductor. How is this different from conduction?*

The third method of charging is by **induction.** Induction involves a rearrangement of electric charges. For induction to occur, a neutral object need only come close to a charged object. No contact is necessary. For example, a negatively charged rubber rod can pick up tiny pieces of paper by induction. The electric charges in the paper are rearranged by the approach of the negatively charged rubber rod. The electrons in the area of the paper nearest to the negatively charged rod are repelled, leaving the positive charges near the rod. Because the positive charges are closer to the negative rod, the paper is attracted. Does this description sound familiar? What method of charging made the wall positive in the area nearest the balloon?

The transfer of electrons from one object to another without further movement is called **static electricity.** The word static means not moving, or stationary. **Static electricity is the buildup of electric charges on an object.** The electric charges build up because electrons have moved from one object to another. However, once built up, the charges do not flow. They remain at rest.

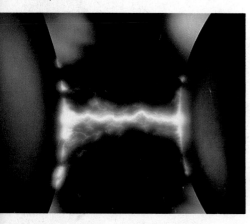

Figure 1-9 *The discharge of static electricity from one metal object to another can be seen as a spark.*

Electric Discharge—Lightning

Electrons that move from one object to another and cause the buildup of charges at rest, or static electricity, eventually leave the object. Sometimes they move onto another object. Usually, these extra electrons escape onto water molecules in the air. (This is why static electricity is much more noticeable on dry days. On dry days the air contains fewer water molecules. Objects are more easily charged because charges cannot escape into the air.) When the charged object loses its static electricity, it becomes neutral once again. The balloon eventually falls off the wall because it loses its charge and there is no longer a force of attraction between it and the wall.

The loss of static electricity as electric charges move off an object is called **electric discharge.** Sometimes electric discharge is slow and quiet. Sometimes it is very rapid and accompanied by a shock, a spark of light, or a crackle of noise.

One of the most dramatic examples of the discharge of static electricity is lightning. During a storm, particles contained in clouds are moved about by the wind. Charges may become separated, and there are buildups of positive and negative charges

Figure 1–10 *Lightning is a spectacular discharge of static electricity between two areas of different charge. Lightning can occur between a portion of a cloud and the ground, between different clouds, or between different parts of the same cloud. Benjamin Franklin's famous experiments provided evidence that lightning is a form of static electricity.*

in different parts of the cloud. If a negatively charged edge of a cloud passes near the surface of the Earth, objects on the Earth become electrically charged by induction. Negative charges move away from the cloud, and positive charges are left closest to the cloud. Soon electrons jump from the cloud to the Earth. The result of this transfer of electrons is a giant spark called lightning.

Lightning can also occur as electrons jump from cloud to cloud. As electrons jump through the air, they produce intense light and heat. The light is the bolt of lightning you see. The heat causes the air to expand suddenly. The rapid expansion of the air is the thunder you hear.

One of the first people to understand lightning as a form of electricity was Benjamin Franklin. In the mid-1700s, Franklin performed experiments that provided evidence that lightning is a form of static electricity, that electricity moves quickly through certain materials, and that a pointed surface attracts electricity more than a flat surface. Franklin suggested that pointed metal rods be placed above the roofs of buildings as protection from lightning. These rods were the first lightning rods.

Lightning rods work according to a principle called grounding. The term grounding comes from the fact that the Earth (the ground) is extremely large and is a good conductor of electric charge.

Figure 1–11 *Lightning rods, such as this one on the Canadian National Tower, provide a safe path for lightning directly into the ground. Scientists studying lightning build up large amounts of electric charge in order to create their own lightning.*

The Earth can easily accept or give up electric charges. Objects in electric contact with the Earth are said to be grounded. A discharge of static electricity usually takes the easiest path from one object to another. So lightning rods are attached to the tops of buildings, and a wire connects the lightning rod to the ground. When lightning strikes the rod, which is taller than the building, it travels through the rod and the wire harmlessly into the Earth. Why is it dangerous to carry an umbrella during a lightning storm?

Unfortunately, other tall objects, such as trees, can also act as grounders. That is why it is not a good idea to stand near or under a tree during a lightning storm. Why do you think it is dangerous to be on a golf course during a lightning storm?

The Electroscope

An electric charge can be detected by an instrument called an **electroscope.** A typical electroscope consists of a metal rod with a knob at the top and a pair of thin metal leaves at the bottom. The rod is inserted into a one-hole rubber stopper that fits into a flask. The flask contains the lower part of the rod and the metal leaves. See Figure 1–12.

In an uncharged electroscope, the leaves hang straight down. When a negatively charged object

FIND OUT BY

DOING

Observing Static Electricity

   1. Place two books about 10 centimeters apart on a table.

   2. Cut tiny paper dolls or some other objects out of tissue paper and place them on the table between the books.

   3. Place a 20- to 25-centimeter-square piece of glass on the books so that the glass covers the paper dolls.

   4. Using a piece of silk, rub the glass vigorously. Observe what happens.

   ■ Using the rule of electric charges and your knowledge of static electricity, explain what you observed.

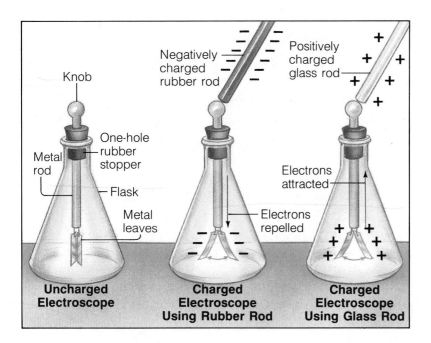

Knob

Metal rod

One-hole rubber stopper

Flask

Metal leaves

Uncharged Electroscope

Negatively charged rubber rod

Electrons repelled

Charged Electroscope Using Rubber Rod

Positively charged glass rod

Electrons attracted

Charged Electroscope Using Glass Rod

Figure 1–12 *An electroscope is used to detect electric charges. Why do the leaves in the electroscope move apart when either a negatively charged rubber rod or a positively charged glass rod makes contact?*

touches the metal knob, electric charges travel down the rod and into the leaves. The leaves spread apart, indicating the presence of an electric charge. Since the charge on both leaves is the same, the leaves repel each other and spread apart.

If a positively charged object touches the knob of the electroscope, free electrons in the leaves and metal rod are attracted by the positive object. The loss of electrons causes the leaves to become positively charged. Again, they repel each other and spread apart.

1–2 Section Review

1. What are three ways an object can acquire an electric charge?
2. What is static electricity?
3. What is electric discharge? Give an example.
4. If the body of a kangaroo contains millions of charged particles, why aren't different kangaroos electrically attracted to or repelled by one another?

Critical Thinking—*Making Inferences*
5. What would happen if a lightning rod were made of an insulating material rather than of a conducting material?

Guide for Reading

Focus on these questions as you read.

▶ *How can a flow of charges be produced?*

▶ *What is the relationship among electric current, voltage, and resistance?*

1–3 The Flow of Electricity

The electricity that you use when you plug an electrical appliance into a wall outlet is certainly not static electricity. If it were, the appliance would not run for long. Useful electricity involves the continuing motion of electric charges. Electric charges can be made to continue flowing.

Producing a Flow of Electrons

A continuing flow of electric charges is produced by a device that changes other forms of energy into electrical energy. Energy is defined as the ability to do work. You use energy when you throw a ball, lift a suitcase, or pedal a bicycle. Energy is involved when a deer runs, a bomb explodes, snow falls—or when electricity flows through a wire. Although there are different types of energy, each type can be converted into any one of the others. A device that converts other forms of energy into electrical energy is known as a source of electricity. Batteries and electric generators are the main sources of electricity. But thermocouples and photocells are other important sources. Electric generators will be discussed in Chapter 3.

BATTERIES A **battery** is a device that produces electricity by converting chemical energy into electrical energy. A battery is made of several smaller units called electric cells, or electrochemical cells. Each cell consists of two different materials called electrodes as well as an electrolyte. The electrolyte is a mixture of chemicals that produces a chemical reaction. The chemical reaction releases electric charges.

Electric cells can be either dry cells or wet cells, depending on the type of electrolyte used. In a wet cell, such as a car battery, the electrolyte is a liquid. In a dry cell, such as the battery in a flashlight, the electrolyte is a pastelike mixture.

Figure 1–14 will help you to understand how a simple electrochemical cell works. In this cell, one of the electrodes is made of carbon and the other is made of zinc. The part of the electrode that sticks up is called the terminal. The electrolyte is sulfuric acid. The acid attacks the zinc and dissolves it.

Figure 1–13 *Imagine plugging a car into the nearest outlet rather than going to the gas station! Researchers have been attempting to design efficient cars that use a rechargeable battery as a source of power.*

Figure 1–14 *Electrochemical cells, which include dry cells and wet cells, convert chemical energy into electric energy. What is a series of dry cells called? What is an example of a wet cell?*

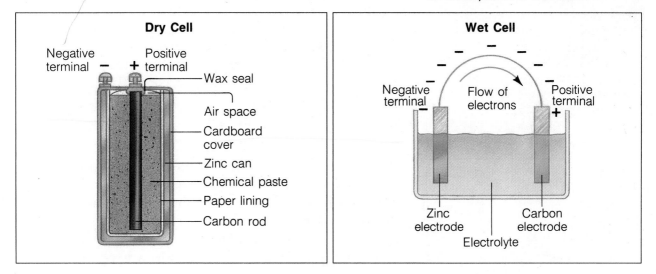

In this process, electrons are left behind on the zinc electrode. Thus the zinc electrode becomes negatively charged. At the same time, a series of chemical reactions causes electrons to be pulled off the carbon electrode. The carbon electrode becomes positively charged. Because there are opposite charges on the electrodes, charge will flow between the terminals if a wire connects them. The difference in charge is called a **potential difference.** A potential difference is like a hill. If a ball is placed at the top of the hill and is allowed to roll down, it will. The steeper the hill, the faster the ball will roll. Similarly, if a potential difference exists between two terminals and a wire connects the terminals, charge will flow. The greater the potential difference, the faster the charge will flow. If the ball is at the bottom of the hill, work will have to be done to roll it up the hill. Work must also be done to move a charge against a potential difference.

THERMOCOUPLES A **thermocouple** is a device that produces electrical energy from heat energy. A thermocouple releases electric charges as a result of temperature differences. In this device the ends of two different metal wires, such as copper and iron, are joined together to form a loop. If one iron-copper junction is heated while the other is cooled, electric charges will flow. The greater the temperature difference between the junctions, the faster the charges will flow. Figure 1–15 shows a thermocouple.

Thermocouples are used as thermometers in cars to show engine temperature. One end of the thermocouple is placed in the engine, while the other end is kept outside the engine. As the engine gets warm, the temperature difference produces a flow of charge. The warmer the engine, the greater the temperature difference—and the greater the flow of charge. The moving charges in turn operate a gauge that shows engine temperature. Thermocouples are also used in ovens and in gas furnaces.

PHOTOCELLS The most direct conversion of energy occurs in a device known as a **photocell.** A photocell takes advantage of the fact that when light with a certain amount of energy shines on a metal surface, electrons are emitted from the surface. These electrons can be gathered in a wire to create a constant flow of electric charge.

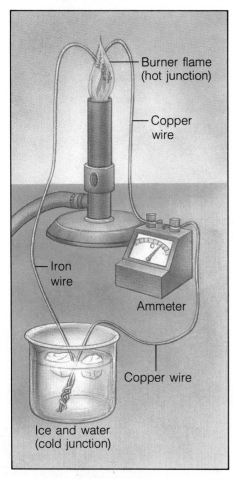

Figure 1–15 *The temperature difference between the hot junction and the cold junction in a thermocouple generates electricity. What is the energy conversion involved in the operation of a thermocouple?*

Burner flame (hot junction)

Copper wire

Iron wire

Ammeter

Copper wire

Ice and water (cold junction)

Figure 1–16 *On November 9, 1965, a major blackout plunged the illuminated skyline of New York City into darkness and left more than 30 million people in the Northeast without electricity.*

Electric Current

When a wire is connected to the terminals of a source, a complete path called a **circuit** is formed. Charge can flow through a circuit. A flow of charge is called an electric **current.** More precisely, electric current is the rate at which charge passes a given point. The higher the electric current in a wire, the faster the electric charges are passing through.

The symbol for current is the letter I. And the unit in which current is expressed is the ampere (A). The ampere, or amp for short, is the amount of current that flows past a point per second. Scientists use instruments such as ammeters and galvanometers to measure current.

You may wonder how charge can flow through a wire. Recall that conductors are made of elements whose atoms have some loosely held electrons. When a wire is connected to the terminals of a source, the potential difference causes the loose electrons to be pulled away from their atoms and to flow through the material.

You may also wonder how lights and other electrical appliances can go on as soon as you turn the switch even though the power plant may be quite a distance away. The answer is that you do not have to wait for the electrons at the power plant to reach your switch. All the electrons in the circuit flow as soon as the switch is turned. To help understand this concept, imagine that each student in Figure 1–17

FIND OUT BY READING

Life on the Prairie

Not very long ago, people just like you grew up without the electrical devices that make your life easy, comfortable, and entertaining. In *Little House on the Prairie* and its related books, Laura Ingalls Wilder tells delightful stories about growing up in the days before electricity. These stories are especially wonderful because they are not tainted by fictional drama. The author simply describes life as it actually was.

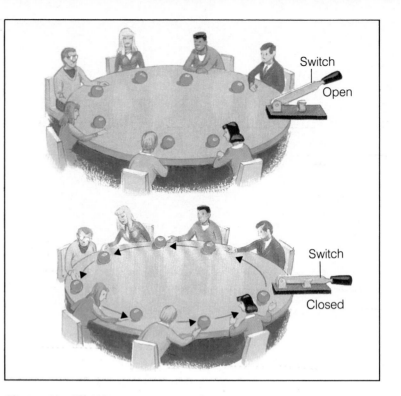

Figure 1–17 *When the switch in the diagram is closed, the students each pass a ball to the right. Thus a ball reaches the switch almost instantly. Similarly, when you flip on a switch to a light, for example, current flows to the light almost immediately.*

CAREERS

Electrical Engineer

The people who design and develop electrical and electronic equipment are called **electrical engineers.** Some electrical engineers specialize in a particular area, such as computer systems, integrated circuitry, or communication equipment. For more information, write to the Institute of Electrical Engineers, U.S. Activities Board, 1111 19th Street, NW, Washington, DC 20036.

has a red ball. When the switch is turned on, each student passes the ball to the person on the right. So almost as soon as the switch is turned on, a red ball reaches the switch. When the switch is turned off, each person still holds a ball—even though it may not be the original ball. This is basically how electrons shift through a conductor. The electrons shift positions, but no electrons leave the circuit.

Voltage

You have already learned that a current flows whenever there is a potential difference between the ends of a wire. The size of the potential difference determines the current that will flow through the wire. The greater the potential difference, the faster the charges will flow. The term **voltage** is often used to describe potential difference. The symbol for voltage is the letter V. Voltage is measured in units called volts (V). If you see the marking 12 V, you know that it means twelve volts. An instrument called a voltmeter is used to measure voltage.

Voltage is not limited to the electrical wires that run your appliances. Voltage, or potential difference, is crucial to your survival. A potential difference exists across the surface of your heart. Changes in the potential difference can be observed on a monitor and recorded as an electrocardiogram (EKG). An EKG is a powerful tool used to locate defects in the heart. Potential differences are also responsible for the movements of your muscles.

In the heart a small region called the pacemaker sends out tiny electrical signals as many as sixty times every minute. This is what causes the heart to beat. Perhaps you may already know that if the heart stops beating, doctors sometimes apply an electric current—potential difference—to the patient's chest. Because the heart is controlled by electricity, it can be made to begin beating again by applying this electric current. If the natural pacemaker fails to work, surgeons can implant an electronic pacemaker.

Humans are not the only organisms that use electricity produced in their bodies. Have you ever heard of electric eels? They truly are electric! An electric eel can produce jolts of electricity up to 650 volts to defend itself or to stun prey. Another type of fish, an electric ray, has a specialized organ in its head that can discharge about 200 volts of electricity to stun and capture prey. Although these voltages may not sound like a lot to you, remember that only 120 volts powers just about everything in your home!

LOW CURRENT AND LOW VOLTAGE

Each electron carries little energy, and there are few electrons. Little total energy is delivered per second.

HIGH CURRENT AND LOW VOLTAGE

Each electron carries little energy, but there are many electrons. Moderate total energy is delivered per second.

LOW CURRENT AND HIGH VOLTAGE

Each electron carries much energy, but there are few electrons. Moderate total energy is delivered per second.

HIGH CURRENT AND HIGH VOLTAGE

Each electron carries much energy, and there are many electrons. High total energy is delivered per second.

Figure 1–18 *This diagram shows the relationship between voltage and current. How is current represented? Voltage?*

Figure 1–19 *Although this interesting-looking fish may appear touchable, such a move could prove to be fatal. This is an electric ray, which, like an electric eel, is capable of producing strong jolts of electricity.*

Resistance

The amount of current that flows through a wire does not depend only on the voltage. It also depends on how the wire resists the flow of electric charge. Opposition to the flow of electric charge is known as **resistance.** The symbol for resistance is the letter R.

Imagine a stream of water flowing down a mountain. Rocks in the stream resist the flow of the water. Or think about running through a crowd of people. The people slow you down by resisting your movement. Electric charges are slowed down by interactions with atoms in a wire. So the resistance of a wire depends on the material of which it is made. If the atoms making up the material are arranged in such a way that it is difficult for electric charge to flow, the resistance of the material will be high. As you might expect, resistance will be less for a wider wire, but more for a longer wire. Higher resistance means greater opposition to the flow of charge. The higher the resistance of a wire, the less current for a given voltage.

The unit of resistance is the ohm (Ω). Different wires have different resistances. Copper wire has less resistance than iron wire does. Copper is a good conductor; iron is a poor conductor. Nonconductors offer such great resistance that almost no current can flow. All electric devices offer some resistance to the flow of current. And although it may not seem so at first, this resistance is often quite useful—indeed, necessary.

You know that light bulbs give off light and heat. Have you ever wondered where the light and heat come from? They are not pouring into the bulb through the wires that lead from the wall. Rather, some of the electric energy passing through the filament of the bulb is converted into light and heat energy. The filament is a very thin piece of metal that resists the flow of electricity within it. The same principle is responsible for toasting your pita bread when you place it in the toaster.

How much resistance a material has depends somewhat on its temperature. The resistance of a metal increases with temperature. At higher temperatures, atoms move around more randomly and thus get in the way of flowing electric charges. At very

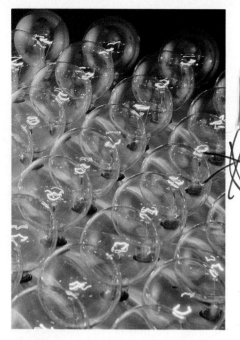

Figure 1–20 *Light bulbs light because of the phenomenon of resistance. The metal filament in the center of the bulb offers enough resistance to the electric current flowing through it so that heat and light are given off.*

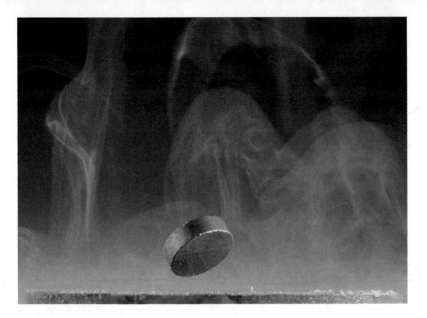

Figure 1–21 *This is no magic trick. At low temperatures, certain materials that have almost no resistance are said to be superconducting. Superconducting materials, such as the one at the bottom of the photograph, repel magnets. For this reason the magnet floats in midair.*

low temperatures, however, the resistance of certain metals becomes essentially zero. Materials in this state are said to be **superconductors.** In superconductors, almost no energy is wasted. However a great deal of energy must be used to keep the material cold enough to be superconducting.

Scientists are currently working to develop new materials that are superconducting at higher temperatures. When this is accomplished, superconductors will become extremely important in industry. Superconductors will be used in large generating plants and in motors where negligible resistance will allow for very large currents. There are also plans for superconducting transmission cables that will reduce energy loss tremendously. Electric generating plants are usually located near major population centers rather than near the fuel source because too much energy is lost in carrying a current. Superconducting transmission lines will make it practical for generating plants to be situated next to fuel sources rather than near population centers. Superconductors are also being tested in high-speed transportation systems.

Ohm's Law

An important expression called **Ohm's Law** identifies the relationship among current, voltage, and resistance. **Ohm's law states that the current in a wire (I) is equal to the voltage (V) divided by the resistance (R).**

CALCULATING

Ohm's Law

Complete the following chart.

I (amps)	V (volts)	R (ohms)
	12	75
15	240	
5.5		20
	6	25
5	110	

As an equation, Ohm's Law is

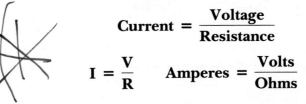

$$\text{Current} = \frac{\text{Voltage}}{\text{Resistance}}$$

$$I = \frac{V}{R} \qquad \text{Amperes} = \frac{\text{Volts}}{\text{Ohms}}$$

If the resistance in a wire is 100 ohms and the voltage is 50 volts, the current is 50/100, or 0.5 ampere. You can rearrange the equation in order to calculate resistance or voltage. What is the resistance if the voltage is 10 volts and the current 2 amperes?

Current Direction

Electrons moving through a wire can move continuously in the same direction, or they can change direction back and forth over and over again.

When electrons always flow in the same direction, the current is called **direct current,** or DC. Electricity from dry cells and batteries is direct current.

When electrons move back and forth, reversing their direction regularly, the current is called **alternating current,** or AC. The electricity in your home is alternating current. In fact, the current in your home changes direction 120 times every second. Although direct current serves many purposes, alternating current is better for transporting the huge amounts of electricity required to meet people's needs.

1–3 Section Review

1. How does an electrochemical cell produce an electric current?
2. What is a thermocouple? A photocell?
3. What is electric current? Explain how current flows through a wire.
4. What is resistance? Voltage? How is electric current related to resistance and voltage?
5. What is direct current? Alternating current?

Critical Thinking—*Drawing Conclusions*
6. If the design of a dry cell keeps electrons flowing steadily, why does a dry cell go "dead"?

1–4 Electric Circuits

Perhaps you wonder why electricity does not flow from the outlets in your home at all times? You can find the answer to this question if you try the following experiment. Connect one wire from a terminal on a dry cell to a small flashlight bulb. Does anything happen? Now connect another wire from the bulb to the other terminal on the dry cell. What happens? With just one wire connected, the bulb will not light. But with two wires providing a path for the flow of electrons, the bulb lights up.

In order to flow, electrons need a closed path through which to travel. **An electric circuit provides a complete, closed path for an electric current.**

Parts of a Circuit

An electric circuit consists of a source of energy; a load, or resistance; wires; and a switch. Recall that the source of energy can be a battery, a thermocouple, a photocell, or an electric generator at a power plant.

The load is the device that uses the electric energy. The load can be a light bulb, an appliance, a machine, or a motor. In all cases the load offers some resistance to the flow of electrons. As a result, electric energy is converted into heat, light, or mechanical energy.

Guide for Reading

Focus on these questions as you read.

▶ *What is an electric circuit?*

▶ *What is the difference between series and parallel circuits?*

Figure 1–22 *No electricity can flow through an open circuit (left). When the switch is flipped, the circuit is closed and electrons have a complete path through which to flow (right). What indicates that a current is flowing through the circuit?*

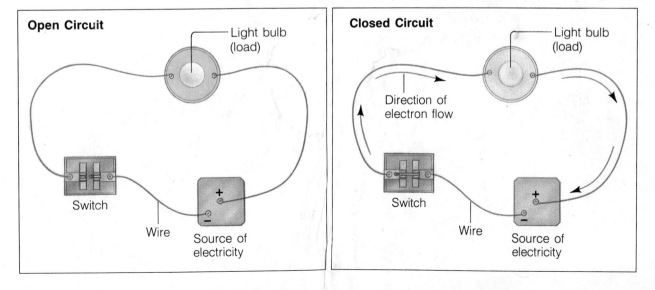

Open Circuit

Light bulb (load)

Switch

Wire

Source of electricity

Closed Circuit

Light bulb (load)

Direction of electron flow

Switch

Wire

Source of electricity

Figure 1–23 *When severe weather conditions—such as the tornado that caused this destruction—damage power lines, the flow of electricity is interrupted. Why?*

The switch in an electric circuit opens and closes the circuit. You will remember that electrons cannot flow through a broken path. Electrons must have a closed path through which to travel. When the switch of an electric device is off, the circuit is open and electrons cannot flow. When the switch is on, the circuit is closed and electrons are able to flow. Remember this important rule: *Electricity cannot flow through an open circuit. Electricity can flow only through a closed circuit.*

Series and Parallel Circuits

There are two types of electric circuits. The type depends on how the parts of the circuit (source, load, wires, and switch) are arranged. If all the parts of an electric circuit are connected one after another, the circuit is a **series circuit.** In a series circuit there is only one path for the electrons to take. Figure 1–24 illustrates a series circuit. The disadvantage of a series circuit is that if there is a break in any part of the circuit, the entire circuit is opened and no current can flow. Inexpensive holiday tree lights are often connected in series. What will happen if one light goes out in a circuit such as this?

In a **parallel circuit,** the different parts of an electric circuit are on separate branches. There are several paths for the electrons to take in a parallel circuit. Figure 1–24 shows a parallel circuit. If there is a break in one branch of a parallel circuit, electrons can still move through the other branches. The

Figure 1–24 *A series circuit provides only one path for the flow of electrons. A parallel circuit provides several paths. How are the circuits in your home wired? Why?*

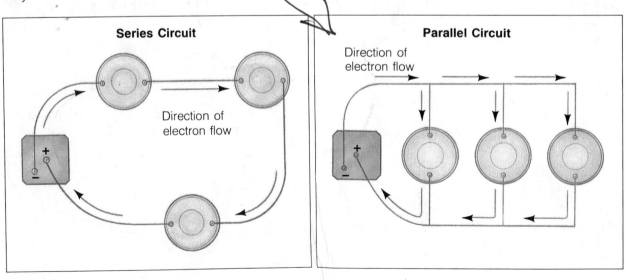

current continues to flow. Why do tree lights connected in parallel have an advantage over tree lights connected in series? Why do you think the electric circuits in your home are parallel circuits?

Household Circuits

Have you ever wondered what was behind the outlet in the wall of your home? After all, it is rather amazing that by inserting a plug into the wall outlet, you can make your television, refrigerator, vacuum cleaner, hair dryer, or any other electrical appliance operate.

Connected to the outlet is a cable consisting of three wires enclosed in a protective casing. Two of the wires run parallel to each other and have a potential difference of 120 volts between them. The third wire is connected to ground. (Recall that a wire that is grounded provides the shortest direct path for current to travel into the Earth.) For any appliance in your home to operate, it must have one of its terminals connected to the high potential wire and the other terminal connected to the low potential wire. The two prongs of a plug of an appliance are connected to the terminals inside the appliance. When the switch of the appliance is closed, current flows into one prong of the plug, through the appliance, and back into the wall through the other prong of the plug.

Many appliances have a third prong on the plug. This prong is attached to the third wire in the cable, which is connected directly to ground and carries no current. This wire is a safety feature to protect against short circuits. A short circuit is an accidental connection that allows current to take a shorter path around a circuit. A shorter path has less resistance and therefore results in a higher current. If the high-potential wire accidentally touches the metal frame of the appliance, the entire appliance will become part of the circuit and anyone touching the appliance will suffer a shock. The safety wire provides a shorter circuit for the current. Rather than flowing through the appliance, the current will flow directly to ground—thereby protecting anyone who might touch the appliance. Appliances that have a plastic casing do not need this safety feature. Can you explain why?

Figure 1–25 *The outlets in a home are connected in such a way that several may rely on the same switch. A home must have several circuits so that different switches control only certain outlets. What happens when the switch in the diagram is flipped off? What if all the appliances in the home are attached to this circuit?*

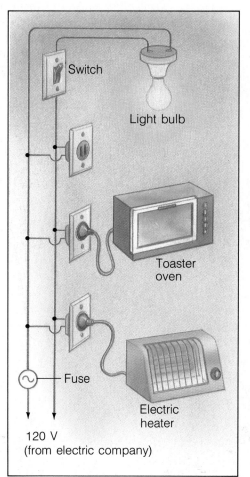

Switch

Light bulb

Toaster oven

Electric heater

Fuse

120 V
(from electric company)

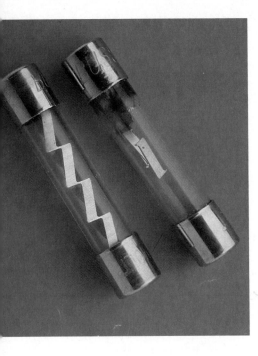

Figure 1-26 *Fuses protect circuits from overloading. The fuse on the left is new. The fuse on the right, however, has been blown and cannot be blown again. How did the blown fuse protect the circuit it was part of?*

Circuit Safety Features

Your home has a great amount of electricity running through it. If too many appliances are running at once on the same circuit or if the wires have become old and frayed, heat can build up in the wiring. If the wires in the walls get too hot, there is the danger of fire. Two devices protect against this potential danger.

FUSES To protect against too much current flowing at once, your home may have **fuses** in a fuse box. Inside each fuse is a thin strip of metal through which current flows. If the current becomes too high, the strip of metal melts and breaks the flow of electricity. So a fuse is an emergency switch.

CIRCUIT BREAKERS One disadvantage of fuses is that once they burn out, they must be replaced. For this reason, **circuit breakers** are often used instead of fuses. Like fuses, circuit breakers protect a circuit from becoming overloaded. Modern circuit breakers have a switch that flips open when the current flow becomes too high. These circuit breakers can easily be reset and used again once the problem has been found and corrected. Circuit breakers are easier to use than fuses.

1-4 Section Review

1. What is an electric circuit?
2. Compare a series circuit and a parallel circuit.
3. Can a circuit be a combination of series connections and parallel connections? Explain your answer.
4. What would happen if your home were not wired in parallel?

Connection—*You and Your World*
5. Does your home have fuses or circuit breakers? Explain the purpose of each device.

PROBLEM Solving

Faulty Wiring

You and your family arrive at the site of your summer vacation—an old but quaint cabin situated at the edge of a beautiful lake. As you pile out of the car, you are greeted by the superintendent responsible for taking care of all the cabins in the area. The neighbors call her Ms. Fix-It.

Ms. Fix-It tells you that everything in the cabin is in working order. However, when she was working on the wiring, she must have made a mistake or two. The kitchen light must remain on in order to keep the refrigerator going. In order to turn on the television, the fan must be on. And the garbage disposal will work only when the oven is on.

Drawing Diagrams

How must the cabin be wired? How should the cabin be rewired? Draw a diagram showing the mistakes and the corrections.

1–5 Electric Power

You probably use the word power in a number of different senses—to mean strength, or force, or energy. To a scientist, **power** is the rate at which work is done or energy is used. **Electric power is a measure of the rate at which electricity does work or provides energy.**

Guide for Reading

Focus on these questions as you read.
▶ What is electric power?
▶ How is electric power related to energy?

POWER USED BY COMMON APPLIANCES	
Appliance	**Power Used (watts)**
Refrigerator/ freezer	600
Dishwasher	2300
Toaster	700
Range/oven	2600
Hair dryer	1000
Color television	300
Microwave oven	1450
Radio	100
Clock	3
Clothes dryer	4000

Figure 1–27 *The table shows the power used by some common appliances. Which appliance would use the greatest amount of electric energy if operated for one hour?*

If you can, pick up a cool light bulb and examine it. Do you notice any words written on it? For example, do you see ''60 watts'' or ''100 watts'' on the bulb? Watts (W) are the units in which electric power is measured. To better understand the meaning of watts, let's look at the concept of electric power more closely.

As you have just read, electric power measures the rate at which electricity does work or provides energy. Electric power can be calculated by using the following equation:

$$\textbf{Power} = \textbf{Voltage} \times \textbf{Current}$$
$$\text{or}$$
$$\textbf{P} = \textbf{V} \times \textbf{I}$$

Or, put another way:

$$\textbf{Watts} = \textbf{Volts} \times \textbf{Amperes}$$

Now think back to the light bulb you looked at. The electricity in your home is 120 volts. The light bulb itself operates at 0.5 ampere. According to the equation for power, multiplying these two numbers gives the bulb's wattage, which in this case is 60 watts. The wattage tells you the power of the bulb, or the rate at which energy is being delivered. As you might expect, the higher the wattage, the brighter the bulb—and the more expensive to run.

To measure large quantities of power, such as the total used in your home, the kilowatt (kW) is used. The prefix *kilo-* means 1000. So one kilowatt is 1000 watts. What is the power in watts of a 0.2-kilowatt light bulb?

Electric Energy

Have you ever noticed the electric meter in your home? This device measures how much energy your household uses. The electric company provides electric power at a certain cost. Their bill for this power is based on the total amount of energy a household uses, which is read from the electric meter.

The total amount of electric energy used depends on the total power used by all the electric appliances

Figure 1–28 *Electricity for your home is purchased on the basis of the amount of energy used and the length of time for which it is used. Power companies install an electric meter in your home to record this usage in kilowatt-hours.*

and the total time they are used. The formula for electric energy is

$$\text{Energy} = \text{Power} \times \text{Time}$$
$$\text{or}$$
$$E = P \times t$$

Electric energy is measured in kilowatt-hours (kWh).

$$\text{Energy} = \text{Power} \times \text{Time}$$
$$\text{Kilowatt-hours} = \text{Kilowatts} \times \text{Hours}$$

One kilowatt-hour is equal to 1000 watts of power used for one hour of time. You can imagine how much power this is by picturing ten 100-watt bulbs in a row, all burning for one hour. One kilowatt-hour would also be equal to a 500-watt appliance running for two hours.

To pay for electricity, the energy used is multiplied by the cost per kilowatt-hour. Suppose the cost of electricity is $0.08 per kilowatt-hour. How much would it cost to burn a 100-watt bulb for five hours? To use a 1000-watt air conditioner for three hours?

Electric Safety

Electricity is one of the most useful energy resources. But electricity can be dangerous if it is not used carefully. Here are some important rules to remember when using electricity.

1. Never handle appliances when your hands are wet or you are standing in water. Water is a fairly good conductor of electricity. If you are wet, you could unwillingly become part of an electric circuit.

2. Never run wires under carpets. Breaks or frays in the wires may go unnoticed. These breaks cause short circuits. A short circuit represents a shorter and easier path for electron flow and thus can cause shocks or a fire.

FIND OUT BY
THINKING

Power and Heat

Examine the appliances in your home for their power rating. Make a chart of this information.

■ What is the relationship between an appliance's power rating and the amount of heat it produces?

Figure 1–29 *Remember to exercise care and good judgment when using electricity. Avoid unsafe conditions such as the ones shown here.*

3. Never overload a circuit by connecting too many appliances to it. Each electric circuit is designed to carry a certain amount of current safely. An overloaded circuit can cause a short circuit.

4. Always repair worn or frayed wires to avoid short circuits.

5. Never stick your fingers in an electric socket or stick a utensil in an appliance that is plugged in. The electricity could be conducted directly into your hand or through the utensil into your hand. Exposure to electricity with both hands can produce a circuit that goes through one arm, across the heart, and out the other arm.

6. Never come close to wires on power poles or to wires that have fallen from power poles or buildings. Such wires often carry very high currents.

FIND OUT BY

CALCULATING

How Much Electricity Do You Use?

1. For a period of several days, keep a record of every electrical appliance you use. Also record the amount of time each appliance is run.

2. Write down the power rating for each appliance you list. The power rating in watts should be marked on the appliance. You can also use information in Figure 1–27.

3. Calculate the amount of electricity in kilowatt-hours that you use each day.

4. Find out how much electricity costs per kilowatt-hour in your area. Calculate the cost of the electricity you use each day.

1–5 Section Review

1. What is electric power? What is the formula for calculating electric power? In what unit is electric power measured?
2. What is electric energy? What is the formula for calculating electric energy? In what unit is electric energy measured?
3. What happens if you touch an exposed electric wire? Why is this situation worse if you are wet or standing in water?

Critical Thinking—*Making Calculations*
4. If left running unused, which appliance would waste more electricity, an iron left on for half an hour or a television left on for one hour?

CONNECTIONS

Electrifying Personalities

Your hair color, eye color, height, and other personal traits are not the haphazard results of chance. Instead, they are determined and controlled by a message found in every one of your body cells. The message is referred to as the *genetic code*. The genetic information passed on from generation to generation in all living things is contained in structures called chromosomes, which are made of genes.

The genetic information contained in a gene is in a molecule of DNA (deoxyribonucleic acid). A DNA molecule consists of a long chain of many small molecules known as nucleotide bases. There are only four types of bases in a DNA molecule: adenine (A), cytosine (C), guanine (G), and thymine (T). The order in which the bases are arranged determines everything about your body.

A chromosome actually consists of two long DNA molecules wrapped around each other in the shape of a double helix. The two strands are held together in a precise shape by electric forces — the attraction of positive charges to negative charges.

In addition to holding the two strands of DNA together, electric forces are responsible for maintaining the genetic code and reproducing it each time a new cell is made. Your body is constantly producing new cells. It is essential that the same genetic message be given to each cell. When DNA is reproduced in the cell, the two strands unwind, leaving the charged parts of the bases exposed. Of the four bases, only certain ones will pair together. A is always paired with T, and G is always paired with C.

Suppose, for example, that after DNA unwinds, a molecule of C is exposed. Of the four bases available for pairing with C, only one will be electrically attracted to C. The charges on the other three bases are not arranged in a way that makes it possible for them to get close enough to those on C.

Electric forces not only hold the two chains together, they also operate to select the bases in proper order during reproduction of the genetic code. Thus the genetic information is passed on accurately to the next generation. So, although it may surprise you, it is a fact: Electricity is partly responsible for your features, from your twinkling eyes to your overall size.

Laboratory Investigation

Electricity From a Lemon

Problem

Can electricity be produced from a lemon, a penny, and a dime?

Materials *(per group)*

bell wire	scissors
cardboard box	sandpaper
compass	dime
lemon	2 pennies

Procedure

1. Wrap 20 turns of bell wire around the cardboard box containing the compass, as shown in the accompanying figure.
2. Roll the lemon back and forth on a table or other flat surface while applying slight pressure. The pressure will break the cellular structure of the lemon.
3. Use the pointed end of the scissors to make two slits 1 cm apart in the lemon.
4. Sandpaper both sides of the dime and two pennies.
5. Insert the pennies in the two slits in the lemon as shown in the figure.

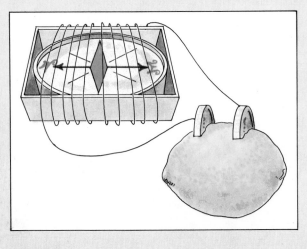

6. Touch the two ends of the bell wire to the coins. Observe any deflection of the compass needle.
7. Replace one of the pennies with the dime. Repeat step 6. Observe any deflection of the compass needle. If there is deflection, observe its direction.
8. Reverse the connecting wires on the coins. Observe any deflection of the compass needle and the direction of deflection.

Observations

1. Is the compass needle deflected when the two ends of the bell wire touch the two pennies?
2. Is the compass needle deflected when the two ends of the bell wire touch the penny and the dime?
3. Is the direction of deflection changed when the connecting wires on the coins are reversed?

Analysis and Conclusions

1. A compass needle will be deflected in the presence of an electric current. Is an electric current produced when two pennies are used? When a dime and a penny are used?
2. What is the purpose of breaking the cellular structure of the lemon? Of sandpapering the coins?
3. What materials are necessary to produce an electric current?
4. An electric current flowing through a wire produces magnetism. Using this fact, explain why a compass is used in this investigation to detect a weak current.
5. A dime is copper with a thin outer coating of silver. What would happen if the dime were sanded so much that the copper were exposed?

Study Guide

Summarizing Key Concepts

1–1 Electric Charge

▲ All matter is made of atoms. Atoms contain positively charged protons, negatively charged electrons, and neutral neutrons.

▲ Opposite charges exert a force of attraction on each other. Similar charges exert a force of repulsion.

1–2 Static Electricity

▲ A neutral object can acquire charge by friction, conduction, or induction.

▲ The buildup of electric charge is called static electricity.

1–3 The Flow of Electricity

▲ Electric charges can be made to flow by a source such as a battery, thermocouple, photocell, or electric generator.

▲ The flow of electrons through a wire is called electric current (I). Electric current is measured in units called amperes (A).

▲ A measure of the potential difference across a source is voltage (V), which is measured in units called volts (V).

▲ Opposition to the flow of charge is called resistance (R). Resistance is measured in units called ohms (Ω).

▲ Ohm's law states that the current in a wire is equal to voltage divided by resistance.

▲ In direct current (DC), electrons flow in one direction. In alternating current (AC), electrons reverse their direction regularly.

1–4 Electric Circuits

▲ An electric circuit provides a complete closed path for an electric current. Electricity can flow only through a closed circuit.

▲ There is only one path for the current in a series circuit. There are several paths in a parallel circuit.

1–5 Electric Power

▲ Electric power measures the rate at which electricity does work or provides energy. The unit of electric power is the watt (W).

Reviewing Key Terms

Define each term in a complete sentence.

1–1 Electric Charge
atom
proton
neutron
electron
charge
force
electric field

1–2 Static Electricity
friction
conduction
conductor
insulator

induction
static electricity
electric discharge
electroscope

1–3 The Flow of Electricity
battery
potential difference
thermocouple
photocell
circuit
current
voltage
resistance

superconductor
Ohm's law
direct current
alternating current

1–4 Electric Circuits
series circuit
parallel circuit
fuse
circuit breaker

1–5 Electric Power
power

Chapter Review

Content Review

Multiple Choice

Choose the letter of the answer that best completes each statement.

1. An atomic particle that carries a negative electric charge is called a(n)
   a. neutron.   c. electron.
   b. positron.   d. proton.
2. Between which particles would an electric force of attraction occur?
   a. proton-proton
   b. electron-electron
   c. neutron-neutron
   d. electron-proton
3. Electricity cannot flow through which of the following?
   a. series circuit   c. parallel circuit
   b. open circuit   d. closed circuit
4. Electric power is measured in
   a. ohms.   c. electron-hours.
   b. watts.   d. volts.

5. The three methods of giving an electric charge to an object are conduction, induction, and
   a. friction.   c. direct current.
   b. resistance.   d. alternating current.
6. Electricity resulting from a buildup of electric charges is
   a. alternating current.
   b. magnetism.
   c. electromagnetism.
   d. static electricity.
7. When electrons move back and forth, reversing their direction regularly, the current is called
   a. direct current.   c. electric charge.
   b. series current.   d. alternating current.

True or False

If the statement is true, write "true." If it is false, change the underlined word or words to make the statement true.

1. The number of electrons in a neutral atom equals the number of <u>protons</u>.
2. A neutral object develops a <u>negative</u> charge when it <u>loses</u> electrons.
3. An instrument that detects charge is a(n) <u>electroscope</u>.
4. Materials that do not allow electrons to flow freely are called <u>insulators</u>.
5. Rubber is a relatively <u>poor</u> conductor of electricity.
6. A <u>photocell</u> generates electricity as a result of temperature differences.
7. Once a <u>circuit breaker</u> burns out, it must be replaced.
8. An electric circuit provides a complete <u>open</u> path for an electric current.
9. Electric <u>power</u> is the rate at which work is done.

Concept Mapping

Complete the following concept map for Section 1–1. Refer to pages P6–P7 to construct a concept map for the entire chapter.

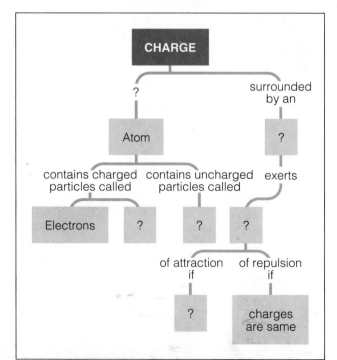

Concept Mastery

Discuss each of the following in a brief paragraph.

1. Describe the structure of an atom. How are atoms related to electric charge?
2. How does the force exerted by a proton on a proton compare with the force exerted by a proton on an electron at the same distance?
3. Describe the three ways in which an object can become charged.
4. Describe how a simple electrochemical cell operates? How are electrochemical cells related to batteries?
5. Compare an insulator and a conductor. How might each be used?
6. Describe two ways in which the resistance of a wire can be increased.
7. What is a circuit? A short circuit?
8. Explain why a tiny 1.5-V cell can operate a calculator for a year, while a much larger 1.5-V cell burns out in a few hours in a toy robot.
9. Discuss three safety rules to follow while using electricity.

Critical Thinking and Problem Solving

Use the skills you have developed in this chapter to answer each of the following.

1. **Making calculations** A light bulb operates at 60 volts and 2 amps.
   a. What is the power of the light bulb?
   b. How much energy does the light bulb need in order to operate for 8 hours?
   c. What is the cost of operating the bulb for 8 hours at a rate of $0.07 per kilowatt-hour?
2. **Identifying relationships** Identify each of the following statements as being a characteristic of (a) a series circuit, (b) a parallel circuit, (c) both a series and a parallel circuit:
   a. $I = V/R$
   b. The total resistance in the circuit is the sum of the individual resistances.
   c. The total current in the circuit is the sum of the current in each resistance.
   d. The current in each part of the circuit is the same.
   e. A break in any part of the circuit causes the current to stop.
3. **Applying concepts** Explain why the third prong from a grounded plug should not be removed to make the plug fit a two-prong outlet.
4. **Making inferences** Electric current can be said to take the path of least resistance. With this in mind, explain why a bird can perch with both feet on a power line and not be injured?

5. **Using the writing process** Imagine that from your window you can see the farm that belongs to your neighbors, whom you have never met. You rarely notice the neighbors, except when it rains. During rainstorms, they protect themselves with huge umbrellas as they walk out to check the crops. Write them a friendly but direct letter explaining why it is dangerous for them to use umbrellas during thunderstorms.

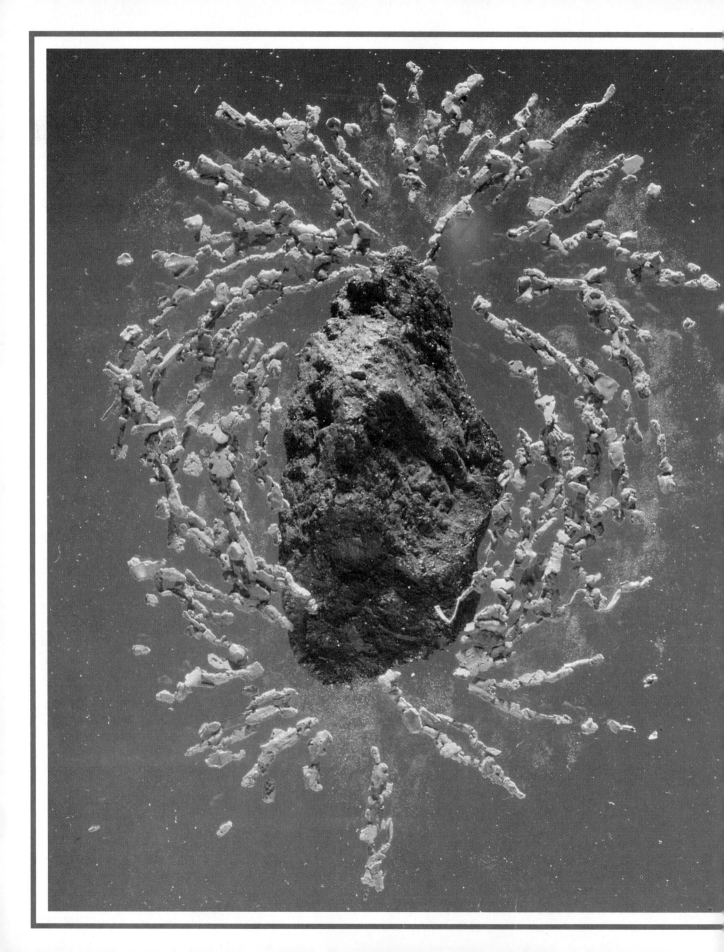

Magnetism

More than 2000 years ago, the Greeks living in a part of Turkey known as Magnesia discovered an unusual rock. The rock attracted materials that contained iron. Because the rock was found in Magnesia, the Greeks named it magnetite. As the Greeks experimented with their new discovery, they observed another interesting thing about this peculiar rock. If they allowed it to swing freely from a string, the same part of the rock would always face in the same direction. That direction was toward a certain northern star, called the leading star, or lodestar. Because of this property, magnetite also became known as lodestone.

The Greeks did not know it then, but they were observing a property of matter called magnetism. In this chapter you will discover what magnetism is, the properties that make a substance magnetic, and the significance of magnetism in your life.

Journal *Activity*

You and Your World You probably use several magnets in the course of a day. In your journal, describe some of the magnets you encounter and how they are used. Also suggest other uses for magnets.

Magnetite, or lodestone, is a natural magnet that exhibits such properties as attracting iron filings.

FIND OUT BY DOING

Magnetic Forces

1. Take two bar magnets of the same size and hold one in each hand.

2. Experiment with the magnets by bringing different combinations of poles together. What do you feel in your hands?

■ Explain your observations in terms of magnetic forces.

2–1 The Nature of Magnets

Have you ever been fascinated by the seemingly mysterious force you feel when you try to push two magnets together or pull them apart? This strange phenomenon is known as **magnetism.** You may not realize it, but magnets play an extremely important role in your world. Do you use a magnet to hold notes on your refrigerator or locker door? Do you play video- and audiotapes? Perhaps you have used the magnet on an electric can opener. Did you know that a magnet keeps the door of your freezer sealed tight? See, you do take advantage of the properties of magnets.

Magnetic Poles

All magnets exhibit certain characteristics. Any magnet, no matter what its shape, has two ends where its magnetic effects are strongest. These regions are referred to as the **poles** of the magnet. One pole is labeled the north pole and the other the south pole. Magnets come in different shapes and sizes. The simplest kind of magnet is a straight bar of iron. Another common magnet is in the shape of a horseshoe. In either case, the poles are at each end. Figure 2–1 shows a variety of common magnets.

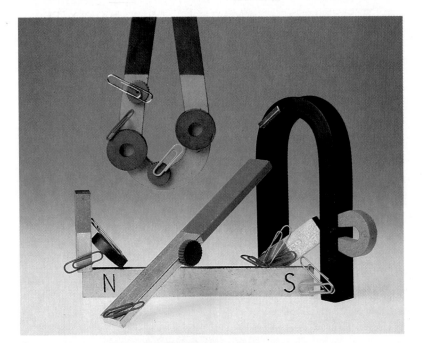

Figure 2–1 *Modern magnets come in a variety of sizes and shapes, including bar magnets, horseshoe magnets, and disc magnets.*

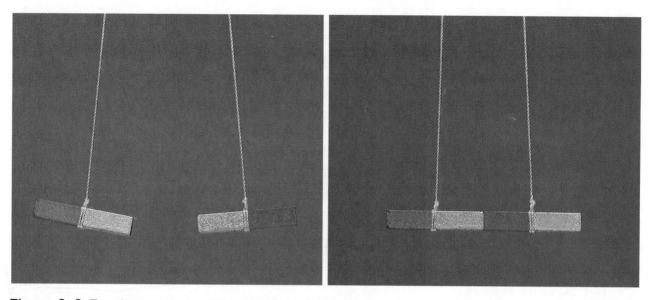

Figure 2-2 *Two bar magnets suspended by strings are free to move. What force is occurring between the magnets in each photograph? Why?*

When two magnets are brought near each other, they exert a force on each other. Magnetic forces, like electric forces, involve attractions and repulsions. If the two north poles are brought close together, they will repel each other. Two south poles do the same thing. However, if the north pole of one magnet is brought near the south pole of another magnet, the poles will attract each other. The rule for magnetic poles is: Like poles repel each other and unlike poles attract each other. How does this rule compare with the rule that describes the behavior of electric charges?

Magnetic poles always appear in pairs—a north pole and a south pole. For many years, physicists have tried to isolate a single magnetic pole. You might think that the most logical approach to separating poles would be to cut a magnet in half. Logical, yes; correct, no. If a magnet is cut in half, two smaller magnets each with a north pole and a south pole are produced. This procedure can be repeated again and again, but a complete magnet is always produced. Theories predict that it should be possible to find a single magnetic pole (monopole), but experimental evidence does not agree. A number of scientists are actively pursuing such a discovery because magnetic monopoles are believed to have played an important role in the early history of the universe.

Figure 2-3 *No matter how many times a magnet is cut in half, each piece retains its magnetic properties. How are magnetic poles different from electric charges?*

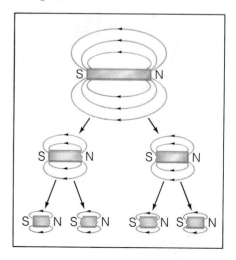

FIND OUT BY DOING

Mapping Lines of Magnetic Force

For this activity you need iron filings, a horseshoe magnet, a thin piece of cardboard, a pencil, and a sheet of paper.

1. Place the horseshoe magnet on a flat surface. Place the cardboard on top of it. Be sure the cardboard covers the entire magnet.

2. Sprinkle iron filings over the cardboard. Make a drawing of the pattern you see.

■ Explain why this particular pattern is formed.

■ What does the pattern tell you about the location of the poles of a horseshoe magnet?

You may think that science has all the answers and that everything has been discovered that can be discovered. But the quest for monopoles illustrates that scientific knowledge is continually developing and changing. It is often the case that a scientific discovery creates a whole new collection of questions to be answered—perhaps by inquisitive minds like yours.

Magnetic Fields

Although magnetic forces are strongest at the poles of a magnet, they are not limited to the poles alone. Magnetic forces are felt around the rest of the magnet as well. The region in which the magnetic forces can act is called a **magnetic field**.

It may help you to think of a magnetic field as an area mapped out by magnetic lines of force. Magnetic lines of force define the magnetic field of an object. Like electric field lines, magnetic field lines can be drawn to show the path of the field. But unlike electric fields, which start and end at charges, magnetic fields neither start nor end. They go around in complete loops from the north pole to the south pole of a magnet. **A magnetic field, represented by lines of force extending from one pole of a magnet to the other, is an area over which the magnetic force is exerted.**

Magnetic lines of force can be easily demonstrated by sprinkling iron filings on a piece of cardboard placed on top of a magnet. See Figure 2–4. Where are the lines of force always the most numerous and closest together?

Figure 2–4 *You can see the magnetic lines of force mapped out by the iron filings placed on a glass sheet above a magnet. The diagram illustrates these lines of force. Where are the lines strongest?*

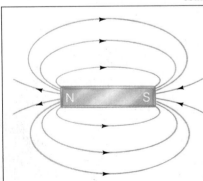

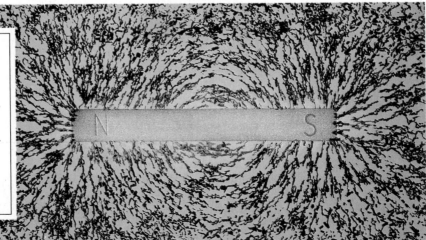

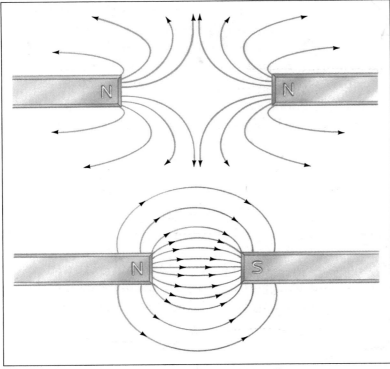

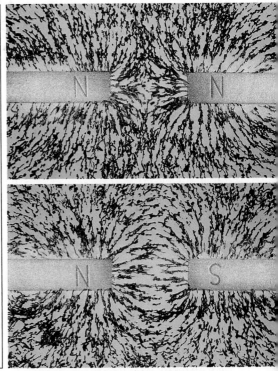

Figure 2–5 *What do the lines of force around these magnets tell you about the interaction of like and unlike magnetic poles?*

Figure 2–5 shows the lines of force that exist between like and unlike poles of two bar magnets. The pattern of iron filings shows that like poles repel each other and unlike poles attract each other.

Magnetic Materials

If you bring a magnet near a piece of wood, glass, aluminum, or plastic, what happens? You are right if you say nothing. There is no action between the magnet and any of these materials. In addition, none of these materials can be magnetized. Yet materials such as iron, steel, nickel, and cobalt react readily to a magnet. And all these materials can be magnetized. Why are some materials magnetic while others are not?

The most highly magnetic materials are called ferromagnetic materials. The name comes from the Latin name for iron, *ferrum.* Ferromagnetic materials are strongly attracted to magnets and can be made into magnets as well. For example, if you bring a strong magnet near an iron nail, the magnet will attract the nail. If you then stroke the nail several times in the same direction with the magnet, the nail itself becomes a magnet. The nail will remain magnetized even after the original magnet is removed.

FIND OUT BY
DOING

Paper Clip Construction

1. How many paper clips can you make stick to the surface of a bar magnet?

■ Explain your results.

2. How many paper clips can you attach in a single row to a bar magnet?

■ Explain your results.

■ What would happen if you placed a plastic-coated paper clip in the second position?

FIND OUT BY DOING

Magnetic Domains

1. Cut several index cards into small strips to represent magnetic domains. Label each strip with a north pole and a south pole.

2. On a sheet of posterboard, arrange the strips to represent an unmagnetized substance.

3. On another sheet of posterboard, arrange the strips to represent a magnetized substance.

Provide a written explanation for your model.

Some materials, such as soft iron, are easy to magnetize. But they also lose their magnetism quickly. Magnets made of these materials are called temporary magnets. Other magnets are made of materials that are more difficult to magnetize, but they tend to stay magnetized. Magnets made of these materials are called permanent magnets. Cobalt, nickel, and iron are materials from which strong permanent magnets can be made. Many permanent magnets are made of a mixture of aluminum, nickel, cobalt, and iron called alnico.

An Explanation of Magnetism

The magnetic properties of a material depend on its atomic structure. Scientists believe that the atom itself has magnetic properties. These magnetic properties are due to the motion of the atom's electrons. Groups of atoms join in such a way that their magnetic fields are all arranged in the same direction, or aligned. This means that all the north poles face in one direction and all the south poles face in the other direction. A region in which the magnetic fields of individual atoms are grouped together is called a **magnetic domain.**

You can think of a magnetic domain as a miniature magnet with a north pole and a south pole. All materials are made up of many domains. In unmagnetized material, the domains are arranged randomly (all pointing in different directions). Because the domains exert magnetic forces in different directions, they cancel out. There is no overall magnetic force in the material. In a magnet, however, most of the domains are aligned. See Figure 2–6.

A magnet can be made from an unmagnetized material such as an iron nail by causing the domains to become aligned. When a ferromagnetic material is placed in a strong magnetic field, the poles of the magnet exert a force on the poles of the individual domains. This causes the domains to shift. Either

Figure 2–6 *The sections represent the various domains of a material. The arrows point toward the north pole of each domain. What is the arrangement of the domains in an unmagnetized material? In a magnetized material?*

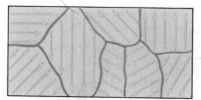

Unmagnetized Material

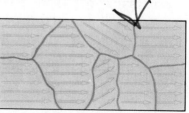

Magnetized Material

most of the domains rotate (turn) to be in the direction of the field, or the domains already aligned with the field become larger while those in other directions become smaller. In both situations, an overall magnetic force is produced. Thus the material becomes a magnet.

This also explains why a magnet can pick up an unmagnetized object, such as a paper clip. The magnet's field causes a slight alignment of the domains in the paper clip so that the clip becomes a temporary magnet. Its north pole faces the south pole of the permanent magnet. Thus it is attracted to the magnet. When the magnet is removed, the domains return to their random arrangement and the paper clip is no longer magnetized.

Even a permanent magnet can become unmagnetized. For example, if you drop a magnet or strike it too hard, you will jar the domains into randomness. This will cause the magnet to lose some or all of its magnetism. Heating a magnet will also destroy its magnetism. This is because the additional energy (in the form of heat) causes the particles of the material to move faster and more randomly. In fact, every material has a certain temperature above which it cannot be made into a magnet at all.

Now that you have learned more about the nature of magnets, you can better describe the phenomenon of magnetism. **Magnetism is the force of attraction or repulsion of a magnetic material due to the arrangement of its atoms—particularly its electrons.**

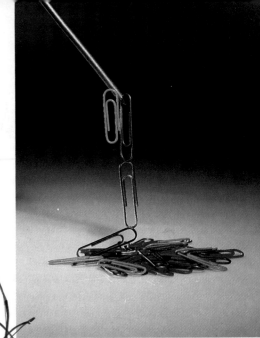

Figure 2–7 *This iron nail attracts metal paper clips. How can an iron nail be turned into a magnet?*

2–1 Section Review

1. How is a magnetic field related to magnetic poles and lines of force?
2. State the rule that describes the behavior of magnetic poles.
3. What is magnetism?
4. What is a magnetic domain? How are magnetic domains related to magnetism?

Critical Thinking—*Applying Concepts*
5. From what you know about the origin of magnetism, explain why cutting a magnet in half produces two magnets.

FIND OUT BY

Where It All Began

The phenomenon of magnetism has been known for many centuries, well before it was actually understood. Various references to this "magical" property can be found in ancient Chinese and Greek mythology. Using reference materials from your library, find out about the discovery and history of lodestones. Write a report describing how lodestone was discovered, what people first thought of it, and how it has come to be understood and used.

PROBLEM Solving

Mix-Up in the Lab

You are working in a research laboratory conducting experiments regarding the characteristics of magnets. You have several samples of magnetic materials and several samples of materials that are not magnetic. Unfortunately, you also have a problem. One of your inexperienced laboratory assistants has removed the label identifying one particular sample as magnetic or not magnetic. To make matters worse, you must complete this part of your research before your boss returns.

Drawing Conclusions

All you have is this photograph showing the pattern of the magnetic domains of the sample. Is the sample a magnet?

Explain how you reached your conclusion. Devise an experiment for your lab assistant to perform to prove your conclusion so that the sample can be correctly labeled.

Guide for Reading

Focus on these questions as you read.

▶ *What are the magnetic properties of the Earth?*

▶ *How does a compass work?*

2–2 The Earth As a Magnet

You have read earlier that as the ancient Greeks experimented with magnetite, they discovered that the same part of the rock always pointed in the same direction. Why does one pole of a bar magnet suspended from a string always point north and the other pole always point south? After all, the poles of a magnet were originally labeled simply to describe the directions they faced with respect to the Earth.

The first person to suggest an answer to this question was an English physician named William Gilbert. In 1600, Gilbert proposed the idea that the Earth itself is a magnet. He predicted that the Earth would be found to have magnetic poles.

Gilbert's theory turned out to be correct. Magnetic poles of the Earth were eventually discovered.

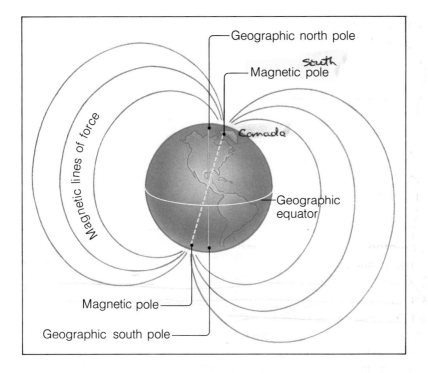

Geographic north pole

south
Magnetic pole

Canada

Geographic equator

Magnetic lines of force

Magnetic pole

Geographic south pole

Figure 2-8 *You can see in this illustration that the magnetic poles are not located exactly at the geographic poles. Does a compass needle, then, point directly north?*

Today, scientists know that the Earth behaves as if it has a huge bar magnet buried deep within it. **The Earth exerts magnetic forces and is surrounded by a magnetic field that is strongest near the north and the south magnetic poles.** The actual origin of the Earth's magnetic field is not completely understood. It is believed to be related to the motion of the Earth's inner core, which is mostly iron and nickel.

Scientists have been able to learn a great deal about the Earth's magnetic field and how it changes over time by studying patterns in magnetic rocks formed long ago. Some minerals have magnetic properties and are affected by the Earth's magnetism. In molten (hot liquid) rocks, the magnetic mineral particles line up in the direction of the Earth's magnetic poles. When the molten rocks harden, a permanent record of the Earth's magnetism remains in the rocks. Scientists have discovered that the history of the Earth's magnetism is recorded in magnetic stripes in the rocks. Although the stripes cannot be seen, they can be detected by special instruments. The pattern of the stripes reveals that the magnetic poles of the Earth have reversed themselves completely many times throughout Earth's history—every half-million years or so.

Figure 2-9 *When volcanic lava hardens into rock, the direction of the Earth's magnetic field at that time is permanently recorded.*

Compasses

If you have ever used a compass, you know that a compass needle always points north. The needle of a compass is magnetized. It has a north pole and a south pole. The Earth's magnetic field exerts a force on the needle just as it exerts a force on a bar magnet hanging from a string.

The north pole of a compass needle points to the North Pole of the Earth. But to exactly which north pole? As you have learned, like poles repel and unlike poles attract. So the magnetic pole of the Earth to which the north pole of a compass points must actually be a magnetic south pole. In other words, the north pole of a compass needle points toward the geographic North Pole, which is actually the magnetic south pole. The same is true of the geographic South Pole, which is actually the magnetic north pole.

The Earth's magnetic poles do not coincide directly with its geographic poles. Scientists have discovered that the magnetic south pole is located in northeastern Canada, about 1500 kilometers from the geographic North Pole. The magnetic north pole is located near the Antarctic Circle. The angular difference between a magnetic pole and a geographic pole is known as magnetic variation, or declination. The extent of magnetic variation is not the same for all places on the Earth. Near the equator, magnetic variation is slight. As you get closer to the poles, the error increases. This must be taken into account when using a compass.

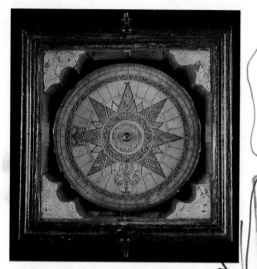

Figure 2–10 *Without the use of compasses, early discoverers would have been unable to chart their courses across the seas and make maps. This photograph shows the earliest surviving Portuguese compass.*

Other Sources of Magnetism in the Solar System

Magnetic fields have been detected repeatedly throughout the galaxy. In addition to Earth, several other planets produce magnetic fields. The magnetic field of Jupiter is ten times greater than that of Earth. Saturn also has a very strong magnetic field. Like Earth, the source of the field is believed to be related to the planet's core.

The sun is another source of a magnetic field. The solar magnetic field extends far above the sun's surface. Streamers of the sun's corona (or outermost

Figure 2–11 *A total solar eclipse provides a glimpse of the sun's corona. The flares of the solar corona are shaped by the sun's magnetic field.*

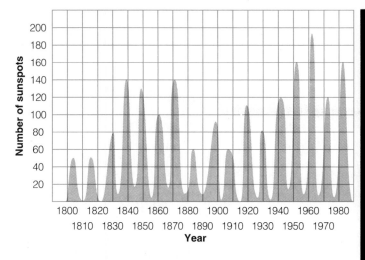

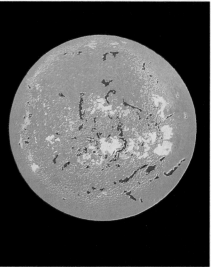

Figure 2–12 *The dark regions on the surface of the sun, or sunspots, are produced by the sun's magnetic field. The pattern of sunspots changes regularly in an 11-year cycle. Notice from the graph how the number of sunspots rises and falls.*

layer) trace the shape of the field. Within specific regions of the sun are very strong magnetic fields. Where magnetic lines of force break through the sun's surface, the temperature of the surface gases is lowered somewhat. These cooler areas appear as dark spots on the surface of the sun. These dark areas are known as sunspots. Sunspots always occur in pairs, each one of the pair representing the opposite poles of a magnet. The annual number of sunspots varies in an eleven-year cycle. The cycle is believed to be related to variations in the sun's magnetic field. Every eleven years the sun's magnetic field reverses, and the north and the south poles switch.

2–2 Section Review

1. In what ways is the Earth like a magnet?
2. How does a compass work?
3. What does it mean to say that the Earth's geographic North Pole is really near the magnetic south pole?
4. What is meant by magnetic declination?
5. How are sunspots related to the sun's magnetic field?

Connection—*Astronomy*

6. If the magnetic field of Earth is related to its inner core, how can astronomers learn about the inner cores of distant planets?

Cork-and-Needle Compass

1. Fill a nonmetal bowl about two-thirds full with water.
2. Magnetize a needle by stroking it with one end of a magnet.
3. Float a cork in the water; then place the needle on the cork. You may need to tape the needle in place.
4. Hold a compass next to the bowl. Compare its needle with the needle on the cork.

■ Explain how the cork-and-needle compass works. What is one disadvantage of a cork-and-needle compass?

2-3 Magnetism in Action

You learned in Chapter 1 that when a charged particle enters an electric field, an electric force is exerted on it (it is either pulled or pushed away). But what happens when a charged particle enters a magnetic field? The magnetic force exerted on the particle, if any, depends on a number of factors, especially the direction in which the particle is moving. **If a charged particle moves in the same direction as a magnetic field, no force is exerted on it. If a charged particle moves at an angle to a magnetic field, the magnetic force acting on it will cause it to move in a spiral around the magnetic field lines.** See Figure 2–14.

SOLAR WIND   The Earth and the other planets are immersed in a wind of charged particles sent out by the sun. These particles sweep through the solar system at speeds of around 400 kilometers per second. If this tremendous amount of radiation (emitted charged particles) reached the Earth, life as we know it could not survive.

Figure 2-13 *Because of its interaction with the solar wind, the Earth's magnetic field differs from that of a bar magnet. The solar wind causes the magnetosphere to stretch out into a tail shape on the side of the Earth that is experiencing nighttime.*

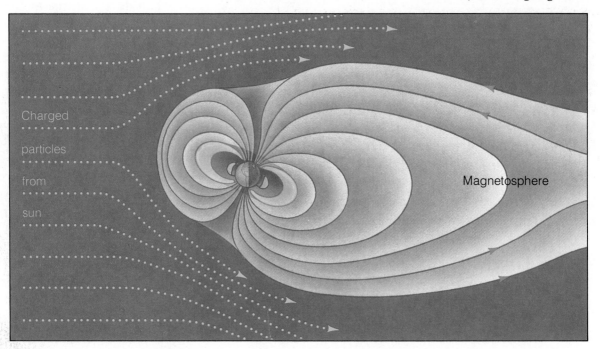

Charged

particles

from

sun

Magnetosphere

Figure 2–14 *Charged particles from the sun become trapped in spiral paths around the Earth's magnetic field lines. What are the two regions in which the particles are confined called?*

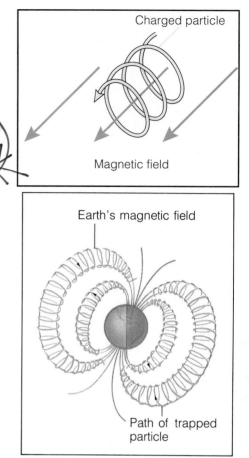

Charged particle

Magnetic field

Earth's magnetic field

Path of trapped particle

Fortunately, however, these charged particles are deflected by the Earth's magnetic field. The magnetic field acts like an obstacle in the path of the solar wind. The region in which the magnetic field of the Earth is found is called the **magnetosphere.** Without the solar wind, the magnetosphere would look like the lines of force surrounding a bar magnet. However, on the side of the Earth facing away from the sun the magnetosphere is blown into a long tail by the solar wind. The solar wind constantly reshapes the magnetosphere as the Earth rotates on its axis.

Sometimes charged particles from the sun do penetrate the field. The particles are forced to continually spiral around the magnetic field lines, traveling back and forth between magnetic poles. Generally, these charged particles are found in two large regions known as the Van Allen radiation belts, named for their discoverer, James Van Allen.

When a large number of these particles get close to the Earth's surface, they interact with atoms in the atmosphere, causing the air to glow. Such a glowing region is called an **aurora.** Auroras are continually seen near the Earth's magnetic poles, since these are the places where the particles are closest to the Earth. The aurora seen in the northern hemisphere is known as the aurora borealis, or northern lights. In the southern hemisphere, the aurora is known as the aurora australis, or southern lights.

Many scientists believe that short-term changes in the Earth's weather are influenced by solar particles and their interaction with the Earth's magnetic field. In addition, during the reversal of the Earth's magnetic field, the field is somewhat weakened. This allows more high-energy particles to reach the Earth's surface. Some scientists hypothesize that these periods during which the magnetic field reversed might have caused the extinction of certain species of plants and animals.

Figure 2–15 *A band of colors called an aurora dances across the sky near the Earth's magnetic poles. This one is in northern Alaska. Why do auroras form?*

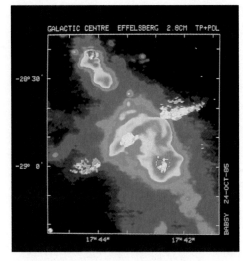

Figure 2–16 *This map of the constellation Sagittarius was made from information collected by a radio telescope.*

FIND OUT BY

A Lonely Voyage

The dangerous quest to reach the North Pole was at one time only a vision in the minds of adventurous explorers. But during the 1800s several different explorers set out on difficult and sometimes fatal missions to the North Pole. Two different explorers reached the pole, but the matter of who arrived first still remains controversial. The answers lie in whether the explorers made appropriate corrections for magnetic declination. Read *To Stand at the Pole: The Dr. Cook-Admiral Peary North Pole Controversy* by William Hunt to discover the details of the story.

ASTRONOMY   Radiation from particles spiraling around in magnetic fields also plays an important role in learning about the universe. Particles trapped by magnetic fields emit energy in the form of radio waves. Radioastronomers study the universe by recording and analyzing radio waves rather than light waves. In this way, scientists have been able to learn a great deal about many regions of the galaxy. For example, energy received from the Crab Nebula indicates that it is a remnant of a supernova (exploding star).

NUCLEAR ENERGY   Utilizing the behavior of charged particles in magnetic fields may be a solution to the energy problem. A tremendous amount of energy can be released when two small atomic nuclei are joined, or fused, together into one larger nucleus. Accomplishing this, however, has been impossible until now because it is extremely difficult to overcome the repulsive electric forces present when two nuclei approach each other. At high temperatures, atoms break up into a gaseous mass of charged particles known as plasma. In this state, nuclear reactions are easier to achieve. But the required temperatures are so high that the charged particles cannot be contained by any existing vessels. By using magnetic fields, however, scientists hope to create a kind of magnetic bottle to contain the particles at the required temperatures. Massive research projects are currently underway to achieve this goal.

2–3 Section Review

1. Describe what happens to a charged particle in a magnetic field.
2. How does the magnetosphere protect the Earth from the sun's damaging radiation?
3. What are Van Allen radiation belts? How are they related to auroras?
4. How might magnetic fields enable scientists to utilize energy from atomic nuclei?

Connection—*Earth Science*
5. Why would a planet close to the sun have to possess a strong magnetic field in order to sustain life?

Magnetic Clues to Giant-Sized Mystery

Have you ever looked at a world map and noticed that the Earth's landmasses look like pieces of a giant jigsaw puzzle? According to many scientists, all the Earth's land was once connected. How then, did the continents form?

In the early 1900s, a scientist named Alfred Wegener proposed that the giant landmass that once existed split apart and its various parts "drifted" to their present positions. His theory, however, met with great opposition because in order for it to be true, it would involve the movement of sections of the solid ocean floor. Conclusive evidence to support his *theory of continental drift* could come only from a detailed study of the ocean floor.

In the 1950s and 1960s, new mapping techniques enabled scientists to discover a large system of underwater mountains, called midocean ridges. These mountains have a deep crack, called a rift valley, running through their center. A great deal of volcanic activity occurs at the midocean ridges. When lava wells up through the rift valley and hardens into rock, new ocean floor is formed. This process is called *ocean-floor spreading.* This evidence showed that the ocean floor could indeed move.

Further evidence came from information about the Earth's magnetic field. When the molten rocks from the midocean ridges harden, a permanent record of the Earth's magnetism remains in the rocks. Scientists found that the pattern of magnetic stripes on one side of the ridge matches the pattern on the other side. The obvious conclusion was that as magma hardens into rocks at the midocean ridge, half the rocks move in one direction and the other half move in the other direction. If it were not for knowledge about magnetism and the Earth's magnetic field, such a conclusion could not have been reached. Yet thanks to magnetic stripes, which provide clear evidence of ocean-floor spreading, the body of scientific knowledge has grown once again.

Laboratory Investigation

Plotting Magnetic Fields

Problem

How can the lines of force surrounding a bar magnet be drawn?

Materials   *(per group)*

bar magnet	horseshoe magnet
sheet of white paper	compass
pencil	

Procedure

1. Place a bar magnet in the center of a sheet of paper. Trace around the magnet with a pencil. Remove the magnet and mark the ends of your drawing to show the north and the south poles. Put the magnet back on the sheet of paper in its outlined position.

2. Draw a mark at a spot about 2 cm beyond the north pole of the magnet. Place a compass on this mark.

3. Note the direction that the compass needle points. On the paper, mark the position of the north pole of the compass needle by drawing a small arrow. The arrow should extend from the mark you made in step 2 and point in the direction of the north pole of the compass needle. This arrow indicates the direction of the magnetic field at the point marked. See the accompanying diagram.

4. Repeat steps 2 and 3 at 20 to 30 points around the magnet. You should have 20 to 30 tiny arrows on the paper when you have finished.

5. Remove the magnet from the paper. Observe the pattern of the magnetic field you have plotted with your arrows.

6. Using a clean sheet of paper, repeat steps 1 through 5 using a horseshoe magnet.

Observations

1. Describe the pattern of arrows you have drawn to represent the magnetic field of the bar magnet. How does it differ from that of the horseshoe magnet?

2. From which pole does the magnetic field emerge on the bar magnet? On the horseshoe magnet?

Analysis and Conclusions

1. What evidence is there that the magnetic field is strongest near the poles of a magnet?

2. What can you conclude about the path of magnetic field lines between the poles of any magnet?

3. **On Your Own**   The pattern of magnetic field lines for a bar magnet will change if another bar magnet is placed near it. Design and complete an investigation similar to this one in which you demonstrate this idea for both like poles and unlike poles.

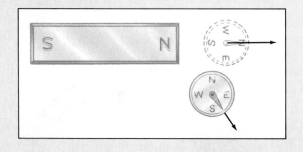

Summarizing Key Concepts

2–1   The Nature of Magnets

▲ Every magnet has two poles—a north pole and a south pole. Like magnetic poles repel each other; unlike poles attract each other.

▲ The region in which magnetic forces can act is called a magnetic field. Magnetic fields are traced by magnetic lines of force.

▲ Magnetic domains are regions in which the magnetic fields of all the atoms line up in the same direction. The magnetic domains of a magnet are aligned. The magnetic domains of unmagnetized material are arranged randomly.

▲ Magnetism is the force of attraction or repulsion exerted by a magnet through its magnetic field.

2–2   The Earth As a Magnet

▲ The Earth is surrounded by a magnetic field that is strongest around the magnetic north and the south poles.

▲ A compass needle does not point exactly to the Earth's geographic poles. It points to the magnetic poles. The difference in the location of the Earth's magnetic and geographic poles is called magnetic declination.

▲ Sunspots on the surface of the sun are related to the sun's magnetic field.

2–3   Magnetism in Action

▲ The deflection of charged particles in a magnetic field is responsible for several phenomena: the protection of the Earth from solar wind; radio astronomy; and future applications of nuclear reactions to produce energy.

▲ The region in which the magnetic field of Earth is exerted is known as the magneto-sphere.

▲ Some charged particles from the sun become trapped by Earth's magnetic field. They are located in two regions known as the Van Allen radiation belts. If charged particles reach Earth's surface, usually at the poles, they create auroras.

Reviewing Key Terms

Define each term in a complete sentence.

2–1   The Nature of Magnets
magnetism
pole
magnetic field
magnetic domain

2–3   Magnetism in Action
magnetosphere
aurora

Chapter Review

Content Review

Multiple Choice

Choose the letter of the answer that best completes each statement.

1. The region in which magnetic forces act is called a
   a. line of force.
   b. pole.
   c. magnetic field.
   d. field of attraction.

2. A region in a magnet in which the magnetic fields of atoms are aligned is a
   a. ferrum.
   b. domain.
   c. compass.
   d. magnetosphere.

3. The idea of the Earth as a magnet was first proposed by
   a. Dalton.
   b. Faraday.
   c. Oersted.
   d. Gilbert.

4. The results of the sun's magnetic field can be seen as
   a. sunspots.
   b. solar winds.
   c. magnetic stripes.
   d. ridges.

5. Which of the following is not a magnetic material?
   a. lodestone
   b. glass
   c. cobalt
   d. nickel

6. The region of the Earth's magnetic field is called the
   a. atmosphere.
   b. stratosphere.
   c. aurora.
   d. magnetosphere.

7. Charged particles from the sun that get close to the Earth's surface produce
   a. supernovas.
   b. volcanoes.
   c. plasma.
   d. auroras.

True or False

If the statement is true, write "true." If it is false, change the underlined word or words to make the statement true.

1. The north pole of a magnet suspended horizontally from a string will point <u>north</u>.
2. Like poles of a magnet <u>attract</u> each other.
3. The region in which magnetic forces act is called a <u>magnetic field</u>.
4. Steel <u>cannot</u> be magnetized.
5. In a magnetized substance, <u>magnetic domains</u> point in the same direction.
6. Magnetic domains exist because of the magnetic fields produced by the motion of <u>electrons</u>.
7. A compass needle points to the Earth's <u>geographic pole</u>.
8. The Earth's magnetic field protects against the harmful radiation in <u>solar winds</u>.

Concept Mapping

Complete the following concept map for Section 2–1. Refer to pages P6–P7 to construct a concept map for the entire chapter.

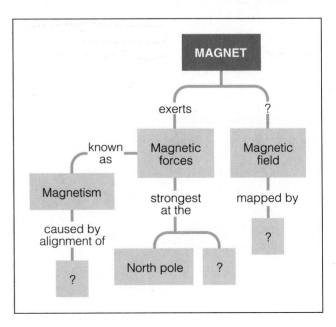

Concept Mastery

Discuss each of the following in a brief paragraph.

1. What are magnetic poles? How do magnetic poles behave when placed next to each other?
2. What happens when a magnet is cut in half? Why?
3. How is a magnetic field represented by magnetic field lines of force?
4. How are magnetic domains related to the atomic structure of a material?
5. How can evidence of changes in the Earth's magnetic field be found in rocks?
6. Why are some materials magnetic while others are not? How can a material be magnetized? How can a magnet lose its magnetism?
7. Like what type of magnet does the Earth act? How?
8. Describe several sources of magnetism in the solar system.
9. Explain how an aurora is produced.
10. What is the significance of the fact that charged particles can become trapped by magnetic fields?

Critical Thinking and Problem Solving

Use the skills you have developed in this chapter to answer each of the following.

1. **Making comparisons** How do the lines of force that arise when north and south poles of a magnet are placed close together compare with the lines of force that arise when two like poles are placed close together? Use a diagram in your explanation.
2. **Applying concepts** Why might an inexperienced explorer using a compass get lost near the geographic poles?
3. **Making comparisons** One proton is traveling at a speed of 100 m/sec parallel to a magnetic field. Another proton is traveling at a speed of 10 m/sec perpendicular to the same magnetic field. Which one experiences the greater magnetic force? Explain.
4. **Identifying patterns** Describe the difference between a permanent magnet and a temporary magnet.
5. **Relating cause and effect** Stroking a material with a strong magnet causes the material to become magnetic. Explain what happens during this process.
6. **Making inferences** Why might a material be placed between two other materials so that magnetic lines of force are not allowed to pass through?
7. **Using the writing process** Write a poem or short story describing the importance of magnets in your daily life. Include at least five detailed examples.

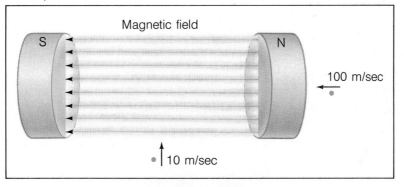

Electromagnetism

Traveling by train can be convenient. But by today's standards, a long train ride can also be time-consuming. The speed of modern trains is limited by the problems associated with the movement of wheels on a track. Engineers all over the world, however, are now involved in the development of trains that "float" above the track. The trains are called maglev trains, which stands for magnetic levitation. Because a maglev train has no wheels, it appears to levitate. While a conventional train has a maximum speed of about 300 kilometers per hour, a maglev train is capable of attaining speeds of nearly 500 kilometers per hour.

Although this may seem to be some sort of magic, it is actually the application of basic principles of electricity and magnetism. Maglev trains are supported and propelled by the interaction of magnets located on the train body and on the track.

In this chapter, you will learn about the intricate and useful relationship between electricity and magnetism. And you will gain an understanding of how magnets can power a train floating above the ground.

Journal *Activity*

You and Your World Have you ever wondered where the electricity you use in your home comes from? In your journal, describe the wires you see connected across poles to all the buildings in your area. If you cannot see them, explain where they must be. Describe some of the factors you think would be involved in providing electricity to an entire city.

A maglev train travels at high speeds without even touching its tracks.

3–1 Magnetism From Electricity

When you think of a source of magnetism, you may envision a bar magnet or a horseshoe magnet. After reading Chapter 2, you may even suggest the Earth or the sun. But would it surprise you to learn that a wire carrying current is also a source of magnetism?

You can prove this to yourself by performing a simple experiment. Bring a compass near a wire carrying an electric current. The best place to hold the compass is just above or below the wire, with its needle parallel to it. Observe what happens to the compass needle when electricity is flowing through the wire and when it is not. What do you observe?

This experiment is similar to one performed more than 150 years ago by the Danish physicist Hans Christian Oersted. His experiment led to an important scientific discovery about the relationship between electricity and magnetism, otherwise known as **electromagnetism.**

Figure 3–1 *A current flowing through a wire creates a magnetic field. This can be seen by the deflection of the compasses around the wire. Notice that the magnetic field is in a circle around the wire. What determines the direction of the magnetic lines of force?*

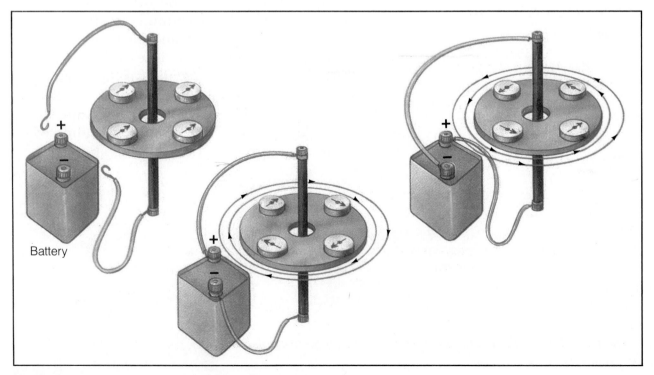

Battery

Oersted's Discovery

For many years, Oersted had believed that a connection between electricity and magnetism had to exist, but he could not find it experimentally. In 1820, however, he finally obtained his evidence. Oersted observed that when a compass is placed near an electric wire, the compass needle deflects, or moves, as soon as current flows through the wire. When the direction of the current is reversed, the needle moves in the opposite direction. When no electricity flows through the wire, the compass needle remains stationary. Since a compass needle is deflected only by a magnetic field, Oersted concluded that an electric current produces a magnetic field. **An electric current flowing through a wire gives rise to a magnetic field whose direction depends on the direction of the current.** The magnetic field lines produced by a current in a straight wire are in the shape of circles with the wire at their center. See Figure 3–1.

Electromagnets

Oersted then realized that if a wire carrying current is twisted into loops, or coiled, the magnetic fields produced by each loop add together. The result is a strong magnetic field in the center and at the two ends, which act like the poles of a magnet. A long coil of wire with many loops is called a **solenoid.** Thus a solenoid acts as a magnet when a current passes through it. The north and the south poles change with the direction of the current.

The magnetic field of a solenoid can be strengthened by increasing the number of coils or the amount of current flowing through the wire. The greatest increase in the strength of the magnetic field, however, is produced by placing a piece of iron in the center of the solenoid. The magnetic field of the solenoid magnetizes—or aligns the magnetic domains of—the iron. The resulting magnetic field is the magnetic field of the wire plus the magnetic field of the iron. This can be hundreds or thousands of times greater than the strength of the field produced by the wire alone. A solenoid with a magnetic material such as iron inside it is called an **electromagnet.**

Making an Electromagnet

Obtain a low-voltage dry cell, nail, and length of thin insulated wire.

1. Remove the insulation from the ends of the wire.

2. Wind the wire tightly around the nail so that you have at least 25 turns.

3. Connect each uninsulated end of the wire to a post on the dry cell.

4. Collect some lightweight metal objects. Touch the nail to each one. What happens?

■ Explain why the device you constructed behaves as it does.

Figure 3–2 *A coil of wire carrying current is a solenoid. As a wire is wound into a solenoid, the magnetic field created by the current becomes strongest at the ends and constant in the center, like that of a bar magnet.*

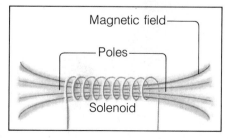

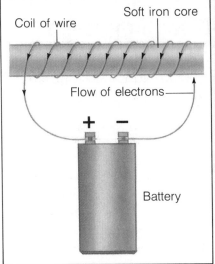

Figure 3–3 *An electromagnet is produced when a piece of soft iron is placed in the center of a solenoid. Large electromagnets can be used to pick up heavy pieces of metal. What factors determine the strength of an electromagnet?*

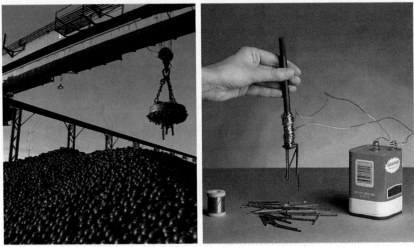

The type of iron used in electromagnets acquires and loses its magnetism when the electric current is turned on and off. This makes electromagnets strong, temporary magnets. Can you think of other ways in which this property of an electromagnet might be useful?

Magnetic Forces on Electric Currents

You have just learned that an electric current exerts a force on a magnet such as a compass. But forces always occur in pairs. Does a magnetic field exert a force on an electric current?

To answer this question, consider the following experiment. A wire is placed in the magnetic field between the poles of a horseshoe magnet. When a current is sent through the wire, the wire jumps up as shown in Figure 3–4. When the direction of the

Figure 3–4 *A magnetic field will exert a force on a wire carrying current. On what does the direction of the force depend?*

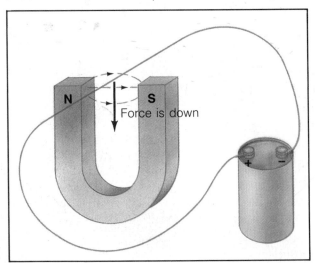

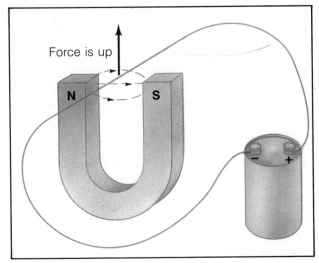

current is reversed, the wire is pulled down. So the answer is yes. **A magnetic field exerts a force on a wire carrying current.** Actually, this fact should not surprise you. After all, as you learned in Chapter 2, a magnetic field can exert a force on a charged particle. An electric current is simply a collection of moving charges.

Applications of Electromagnetism

A number of practical devices use solenoids and electromagnets because they can move mechanical parts quickly and accurately.

ELECTRIC MOTOR An **electric motor** is a device that changes electric energy into mechanical energy that is used to do work. (Mechanical energy is energy related to motion.) An electric motor contains an electromagnet that is free to rotate on a shaft and a permanent magnet that is held in a fixed position. The electromagnet that is free to rotate is called the armature. The armature rotates between the poles of the permanent magnet.

When a current flows through the coil of wire, the ends of the electromagnet become magnetic poles. (Recall that an electromagnet acts as a magnet whose poles depend on the direction of the current.) Because the poles of the armature are next to the same poles on the fixed magnet, they are repelled. The armature begins to rotate toward the opposite poles. But just before the armature reaches the poles, the current through the wire changes direction. This causes the poles of the armature to switch. Once again the poles of the armature and the poles of the fixed magnet repel. So the armature continues to rotate. This process occurs over and over, keeping the armature in continuous rotation.

A current that switches direction every time the armature turns halfway is essential to the operation of an electric motor. The easiest way to supply a changing current is to use AC, or alternating current, to run an electric motor. Recall that alternating current is constantly changing direction.

An electric motor can be made to run on DC, or direct current, by using a reversing switch known as a commutator. A commutator is attached to the

FIND OUT BY DOING

Changing Direction

Obtain a dry cell, a compass, and a length of thin insulated wire.

1. Remove the insulation from both ends of the wire.

2. Place the compass flat on the table. Observe the direction of the needle.

3. Connect each uninsulated end of the wire to a post on the dry cell. Put the center of the wire across the compass. Observe the needle.

4. Disconnect the wires and reconnect them to the opposite terminals without moving the wire over the compass. What happens? Repeat this procedure.

■ How is the magnetic field produced by an electric current related to the direction of the current?

Figure 3–5 *A motor can be found within most practical devices.*

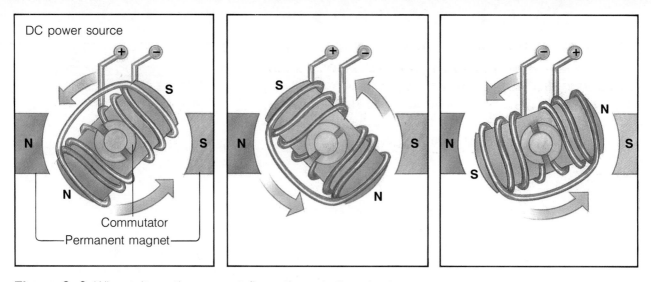

Figure 3–6 *When alternating current flows through the wire in a motor, the poles of the movable electromagnet are reversed. The alternating attraction and repulsion between this electromagnet and a stationary magnet cause the movable electromagnet to spin on its shaft. Thus electric energy is converted into mechanical energy.*

movable electromagnet (or armature). As the electromagnet turns, the commutator switches the direction of current so that the magnetic poles of the electromagnet reverse and the electromagnet spins. Electric current is supplied to the commutator through contacts called brushes. The brushes do not move; they simply touch the commutator as it spins.

The mechanical energy of the spinning armature turns the shaft of the motor, enabling the motor to do work. The shaft can be connected to pulleys, fan blades, wheels, or almost any other device that uses mechanical energy. Sewing machines, refrigerators, vacuum cleaners, power saws, subway trains, and most industrial machinery are a few of the many devices that would be practically impossible to design without electric motors.

Figure 3–7 *As current flows through the wire in a galvanometer, the magnet forces the wire to turn and deflect the needle attached to it. The greater the current, the greater the deflection.*

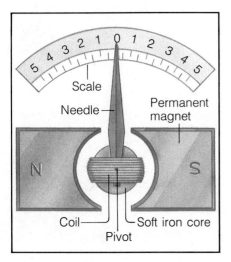

GALVANOMETER Another instrument that depends on electromagnetism is a **galvanometer.** A galvanometer is an instrument used to detect small currents. It is the basic component of many meters, including the ammeter and voltmeter you read about in Chapter 1. A galvanometer consists of a coil of wire connected to an electric circuit and a needle. The wire is suspended in the magnetic field of a permanent magnet. When current flows through the wire, the magnetic field exerts a force on the wire, causing it to move the needle of the galvanometer. The size of the current will determine the amount of force on the wire, and the amount the needle will move. Because the needle will move in the opposite direction when the current is reversed, a galvanometer can be used to measure the direction of current as well as the amount.

OTHER COMMON USES A simple type of electromagnet consists of a solenoid in which an iron rod is only partially inserted. Many doorbells operate with this type of device. See Figure 3–8. When the doorbell is pushed, the circuit is closed and current flows through the solenoid. The current causes the solenoid to exert a magnetic force. The magnetic force pulls the iron into the solenoid until it strikes a bell.

The same basic principle is used in the starter of an automobile. When the ignition key is turned, a circuit is closed, causing an iron rod to be pulled into a solenoid located in the car's starter. The movement of the iron rod connects the starter to other parts of the engine and moves a gear that enables the engine to turn on. Another example of the application of this principle can be found in a washing machine. The valves that control the flow of water into the machine are opened and closed by the action of an iron rod moving into a solenoid.

Some other uses of electromagnets include telephones, telegraphs, and switches in devices such as tape recorders. Electromagnets also are important in heavy machinery that is used to move materials, such as moving scrap metal from one place to another.

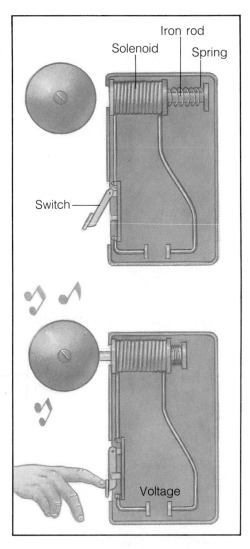

Figure 3–8 *Whenever you ring a doorbell, you are using a solenoid. When the button is pushed, current flows through the wire causing the solenoid to act as a magnet. Once magnetic, the solenoid attracts the bar of iron until the iron strikes the bell, which then rings.*

3–1 Section Review

1. How is magnetism related to electricity?
2. What is an electromagnet? What are some uses of electromagnets?
3. How is an electromagnet different from a permanent magnet?
4. How does an electromagnet change electric energy to mechanical energy in an electric motor?

Critical Thinking—*Making Comparisons*
5. How is the effect of an electric current on a compass needle different from the effect of the Earth's magnetic field on a compass needle?

3–2 Electricity From Magnetism

If magnetism can be produced from electricity, can electricity be produced from magnetism? Scientists who learned of Oersted's discovery asked this very question. In 1831, the English scientist Michael Faraday and the American scientist Joseph Henry independently provided the answer. It is interesting to note that historically Henry was the first to make the discovery. But because Faraday published his results first and investigated the subject in more detail, his work is better known.

Electromagnetic Induction

In his attempt to produce an electric current from a magnetic field, Faraday used an apparatus similar to the one shown in Figure 3–9. The coil of wire on the left is connected to a battery. When current flows through the wire, a magnetic field is produced. The strength of the magnetic field is increased by the iron, as in an electromagnet. Faraday hoped that the steady current would produce a magnetic field strong enough to create a current in the wire on the right. But no matter how strong the current he used, Faraday could not achieve his

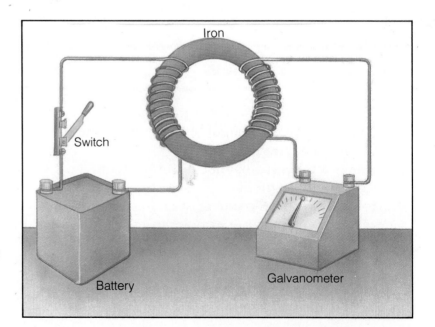

Figure 3–9 *Using this setup, Faraday found that whenever the current in the wire on the left changed, a current was induced in the wire on the right. The changing current produced a changing magnetic field, which in turn gave rise to a current.*

Iron

Switch

Battery

Galvanometer

desired results. The magnetic field did not produce a current in the second wire. However, something rather strange caught Faraday's attention. The needle of the galvanometer deflected whenever the current was turned on or off. Thus a current was produced in the wire on the right, but only when the current (and thus the magnetic field) was changing.

Faraday concluded that although a steady magnetic field produced no electric current, a changing magnetic field did. Such a current is called an **induced current.** The process by which a current is produced by a changing magnetic field is called **electromagnetic induction.**

Faraday did many other experiments into the nature of electromagnetic induction. In one, he tried moving a magnet near a closed loop of wire. What he found out was that when the magnet is held still, there is no current in the wire. But when the magnet is moved, a current is induced in the wire. The direction of the current depends on the direction of movement of the magnet. In another experiment he tried holding the magnet still while moving the wire circuit. In this case, a current is again induced.

The one common element in all Faraday's experiments is a changing magnetic field. It does not matter whether the magnetic field changes because the magnet moves, the circuit moves, or the current

FIND OUT BY **DOING**

Compass Interference

Use a compass to explore the magnetic properties of a room. Take the compass to different parts of the room to see which way the needle points. Can you find places where the compass does not point north? Why does it point in different directions?

■ What precautions must a ship's navigator take when using a compass?

Figure 3–10 *A current is induced in a wire that is exposed to a changing magnetic field. Here the magnet is moved past a stationary wire. On what does the direction of the current depend?*

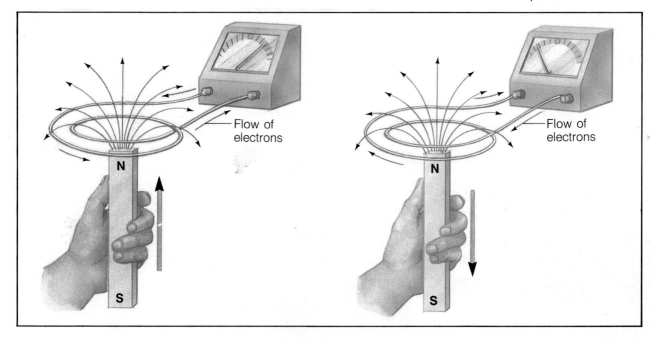

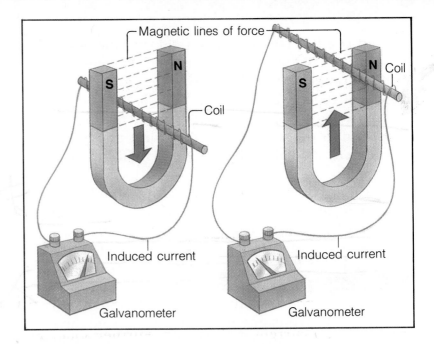

Figure 3–11 *A current is also induced in a wire when the wire is moved through a stationary magnetic field.*

F IND OUT BY WRITING

The History of Electromagnetism

Several scientists were responsible for establishing the relationship between electricity and magnetism. Using books and reference materials in the library, write a report about the scientists listed below. Include information about their lives as well as their contributions to a better understanding of electromagnetism.

Hans Christian Oersted
André Ampère
Michael Faraday
Joseph Henry
Nikola Tesla

giving rise to the magnetic field changes. It matters only that a changing magnetic field is experienced. **An electric current will be induced in a circuit exposed to a changing magnetic field.**

It may help you to think about electromagnetic induction in terms of magnetic lines of force. In each case magnetic lines of force are being cut by a wire. When a conducting wire cuts across magnetic lines of force, a current is produced.

GENERATORS An important application of electromagnetic induction is a **generator.** A generator is a device that converts mechanical energy into electrical energy. How does this energy conversion compare with that in an electric motor? Generators in power plants are responsible for producing about 99 percent of the electricity used in the United States.

A simple generator consists of a loop of wire mounted on a rod, or axle, that can rotate. The loop of wire, which is attached to a power source, is placed between the poles of a magnet. When the loop of wire is rotated by the power source, it moves through the field of the magnet. Thus it experiences a changing magnetic field (magnetic lines of force are being cut). The result is an induced current in the wire.

As the loop of wire continues to rotate, the wire moves parallel to the magnetic lines of force. At this point the field is not changing and no lines of force

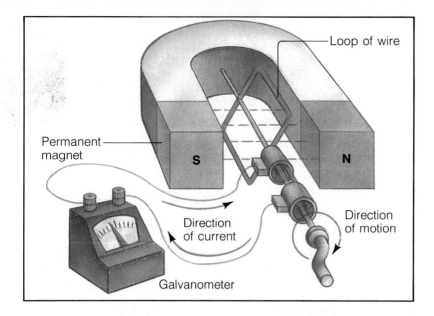

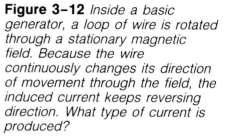

Figure 3–12 *Inside a basic generator, a loop of wire is rotated through a stationary magnetic field. Because the wire continuously changes its direction of movement through the field, the induced current keeps reversing direction. What type of current is produced?*

are cut, so no current is produced. Further rotation moves the loop to a position where magnetic lines of force are cut again. But this time the lines of force are cut from the opposite direction. This means that the induced current is in the opposite direction. Because the direction of the electric current changes with each complete rotation of the wire, the current produced is AC, or alternating current.

The large generators in power plants have many loops of wire rotating inside large electromagnets. The speed of the generators is controlled very carefully. The current is also controlled so that it reverses direction 120 times each second. Because two reversals make one complete cycle of alternating

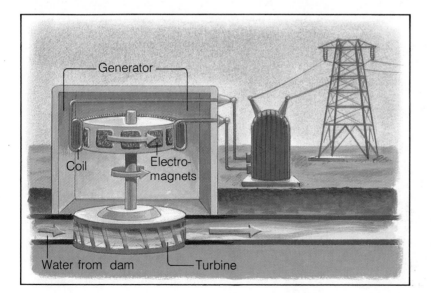

Figure 3–13 *In the operation of a modern generator, a source of mechanical energy—such as water—spins a turbine. The turbine moves large electromagnets encased in coils of insulated wire. As the electromagnets move, the coiled wire cuts through a magnetic field, inducing an electric current in the wire.*

current, the electricity generated has a frequency of 60 hertz. (Frequency is the number of cycles per second, measured in hertz.) Alternating current in the United States has a frequency of 60 hertz. In many other countries, the frequency of alternating current is 50 hertz.

Generators can also be made to produce direct current. In fact, the system proposed and developed by Thomas Edison distributed direct-current electricity. However, the method of alternating current, developed by Edison's rival Nikola Tesla, eventually took its place. When only direct current is used, it has to be produced at high voltages, which are extremely dangerous to use in the home and office. Alternating current, however, can be adjusted to safe voltage levels.

You may wonder about what type of power source is responsible for turning the loop of wire in a generator. In certain generators it can be as simple as a hand crank. But in large generators turbines provide the mechanical energy to turn the axles. Turbines are wheels that are turned by the force of moving wind, water, or steam. Most of the power used in the United States today is generated at steam

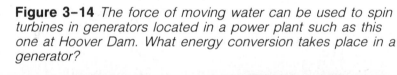

Figure 3–14 *The force of moving water can be used to spin turbines in generators located in a power plant such as this one at Hoover Dam. What energy conversion takes place in a generator?*

plants. There the heat from the burning of fossil fuels (coal, oil, natural gas) or from nuclear fission boils water to produce high-pressure steam that turns the turbines. In many modern generators, the loop remains stationary, but the magnetic field rotates. The magnetic field is produced by a set of electromagnets that rotate.

If you own a bicycle that has a small generator attached to the back wheel to operate the lights, you are the source of power for the generator. To turn the lights on, a knob on the generator is moved so that it touches the wheel. As you pedal the bike, you provide the mechanical energy to turn the wheel. The wheel then turns the knob. The knob is attached to a shaft inside the generator. The shaft rotates a coil of wire through a magnetic field. What happens to the lights when you stop pedaling?

A small electric generator, often called an alternator, is used to recharge the battery in a car while the engine is running. Even smaller electric generators are used extensively in the conversion of information into electric signals. When you play a conventional record, for example, the grooves in the record cause the needle to wiggle. The needle is connected to a tiny magnet mounted inside a coil. As the magnet wiggles, a current is induced that corresponds to the sounds in the record grooves. This signal drives loudspeakers, which themselves use magnetic forces to convert electric signals to sound.

When you play a videotape or an audiotape, you are again taking advantage of induced currents. When information is taped onto a videotape or audiotape, it is in the form of an electric signal. The electric signal is fed to an electromagnet. The strength of the field of the electromagnet depends on the strength of the electric signal. The tape is somewhat magnetic. When it passes by the electromagnet, the magnetic field pulls on the magnetic domains of the tape. According to the electric signal, the domains are arranged in a particular way. When the recorded tape is played, a tiny current is induced due to the changing magnetic field passing the head of the tape player. The magnetic field through the tape head changes with the changing magnetic field of the tape, inducing a current that corresponds to

Figure 3–15 *A light powered by a generator on a bicycle uses the mechanical energy of the spinning tire to rotate the wire. Can the light be on when the bicycle is not in use?*

the information—audio, video, or computer data—recorded on the tape. Magnetic disks used for information storage in computers work on the same principle. For this reason, placing a tape next to a strong magnet or a device consisting of an electromagnet can destroy the information on the tape.

Devices that use electromagnetic induction have a wide variety of applications. In the field of geophysics, a device called a geophone, or seismometer, is used to detect movements of the Earth—especially those associated with earthquakes. A geophone consists of a magnet and a coil of wire. Either the magnet or the wire is fixed to the Earth and thus moves when the Earth moves. The other part of the geophone is suspended by a spring so that it does not move. When the Earth moves, a changing magnetic field is experienced and an electric current is induced. This electric current is converted into an electric signal that can be detected and measured. Why is a geophone a valuable scientific instrument?

TRANSFORMERS   A **transformer** is a device that increases or decreases the voltage of alternating current. A transformer operates on the principle that a current in one coil induces a current in another coil.

A transformer consists of two coils of insulated wire wrapped around the same iron core. One coil is called the primary coil and the other coil is called the secondary coil. When an alternating current passes through the primary coil, a magnetic field is created. The magnetic field varies as a result of the alternating current. Electromagnetic induction causes a current to flow in the secondary coil.

If the number of loops in the primary and the secondary coils is equal, the induced voltage of the secondary coil will be the same as that of the primary coil. However, if there are more loops in the secondary coil than in the primary coil, the voltage of the secondary coil will be greater. Since this type of transformer increases the voltage, it is called a step-up transformer.

In a step-down transformer, there are fewer loops in the secondary coil than there are in the primary coil. So the voltage of the secondary coil is less than that of the primary coil.

Transformers play an important role in the transmission of electricity. Power plants are often situated

An Electrifying Experience

1. Remove the insulation from both ends of a length of thin insulated wire.

2. Coil the wire into at least 7 loops.

3. Connect each uninsulated end of the wire to a terminal of a galvanometer.

4. Move the end of a bar magnet halfway into the loops of the wire. Observe the galvanometer needle.

5. Move the magnet faster and slower. Observe the needle.

6. Increase and decrease the number of loops of the wire. Observe the needle.

7. Move the magnet farther into the loops and not as far. Observe the needle.

8. Use a strong bar magnet and a weak one. Observe the needle.

- What process have you demonstrated?
- Explain all your observations.
- What variables affect the process the most? The least?

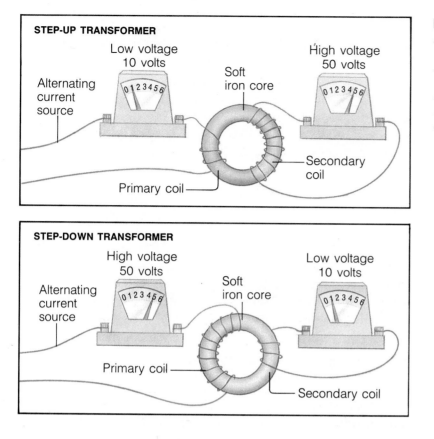

STEP-UP TRANSFORMER

Low voltage 10 volts

Alternating current source

0 1 2 3 4 5 6

Soft iron core

High voltage 50 volts

0 1 2 3 4 5 6

Secondary coil

Primary coil

STEP-DOWN TRANSFORMER

High voltage 50 volts

Alternating current source

0 1 2 3 4 5 6

Soft iron core

Low voltage 10 volts

0 1 2 3 4 5 6

Primary coil

Secondary coil

Figure 3–16 *A transformer either increases or decreases the voltage of alternating current. A step-up transformer increases voltage. A step-down transformer decreases voltage. Which coil has the greater number of loops in each type of transformer?*

long distances from metropolitan areas. As electricity is transmitted over long distances, there is a loss of energy. At higher voltages and lower currents, electricity can be transmitted with less energy wasted. But if power is generated at lower voltages and also used at lower voltages, how can high-voltage transmission be achieved? By stepping up the voltage before transmission (step-up transformer) and stepping it back down before distribution (step-down transformer), power can be conserved.

Step-up transformers are used by power companies to transmit high-voltage electricity to homes and offices. They are also used in fluorescent lights and X-ray machines. In television sets, step-up transformers increase ordinary household voltage from 120 volts to 20,000 volts or more.

Figure 3–17 *Electric current is most efficiently transmitted at high voltages. However, it is produced and used at low voltages. For this reason, electric companies use transformers such as these to adjust the voltage of electric current. Why is alternating current more practical than direct current?*

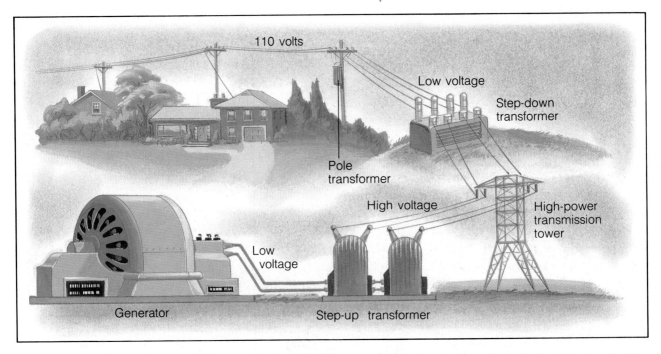

110 volts

Low voltage

Step-down transformer

Pole transformer

High voltage

High-power transmission tower

Low voltage

Generator

Step-up transformer

Figure 3–18 *Low-voltage current produced at a power plant is stepped-up before being sent over long wires. Near its destination, high-voltage current is stepped-down before it is distributed to homes and other buildings. Current may be further transformed in appliances that require specific voltages.*

Step-down transformers reduce the voltage of electricity from a power plant so it can be used in the home. Step-down transformers are also used in doorbells, model electric trains, small radios, tape players, and calculators.

3–2 Section Review

1. What is electromagnetic induction?
2. How can an electric current be produced from a magnetic field?
3. What is the purpose of a generator?
4. What is the difference between a step-up and a step-down transformer?

Connection—*You and Your World*
5. What are some common objects that use either electromagnetism or electromagnetic induction?

CONNECTIONS

Smaller Than Small

When you have difficulty seeing a very small object, do you use a magnifying glass to help you? A magnifying glass enlarges the image of the object you are looking at. If you want to see an even smaller object, you may need a *microscope.* A microscope is an instrument that makes small objects appear larger. The Dutch biologist Anton van Leeuwenhoek is given credit for developing the first microscope. His invention used glass lenses to focus light.

The light microscope, which has descended from Van Leeuwenhoek's original device, still uses lenses in such a way that an object is magnified. Light microscopes are very useful. But due to the properties of light, there is a limit to the magnification light microscopes can achieve. How then can objects requiring greater magnification be seen? In the 1920s, scientists realized that a beam of electrons could be used in much the same way as light was. But microscopes that use electron beams cannot use glass lenses. Instead, they use lenses consisting of magnetic fields that exert forces on the electrons to bring them into focus. The magnetic fields are produced by electromagnets.

Electron microscopes have one major drawback. The specimens to be viewed under an electron microscope must be placed in a vacuum. This means that the specimens can no longer be alive. Despite this disadvantage, electron microscopes are extremely useful for studying small organisms, parts of organisms, or the basic structure of matter. Electron microscopes have opened up a new world!

High-precision microscopes have enabled researchers to peer into the world of the tiny. On this scale you may hardly recognize an ant's head, the cells responsible for pneumonia, or a stylus traveling through the grooves of a record.

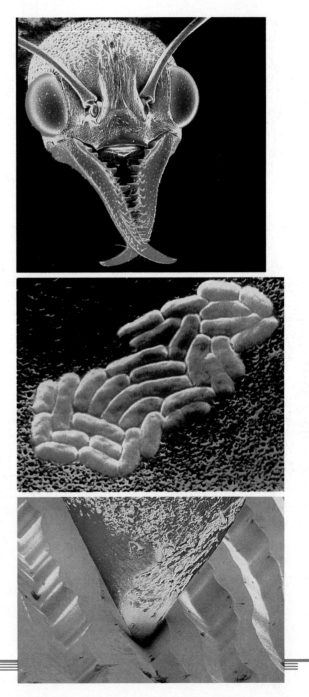

Laboratory Investigation

Electromagnetism

Problem

What factors affect the strength of an electro-magnet? What materials are attracted to an electromagnet?

Materials *(per group)*

dry cell	nickel
6 paper clips	dime
5 iron nails, 10 cm long	
2 meters of bell wire	
small piece of aluminum foil	
penny or copper sheet	
other objects to be tested	

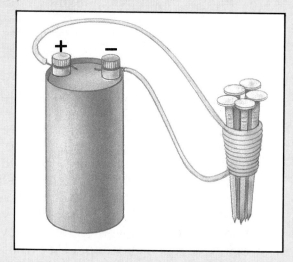

Procedure ⊣▮⊢

1. Hold the five nails together and neatly wrap the wire around them. Do not allow the coils to overlap. Leave about 50 cm of wire at one end and about 100 cm at the other end.

2. Attach the shorter end of the wire to one terminal of the dry cell.

3. Momentarily touch the 100-cm end of the wire to the other terminal of the dry cell. **CAUTION:** *Do not operate the electro-magnet for more than a few seconds each time.*

4. When the electromagnet is on, test each material for magnetic attraction. Record your results.

5. During the time the electromagnet is on, determine the number of paper clips it can hold.

6. Wrap the 100-cm end of wire over the first winding to make a second layer. You should use about 50 cm of wire. There should be approximately 50 cm of wire remaining.

7. Connect the wire once again to the dry cell. Determine the number of paper clips the electromagnet can now hold. Record your results.

8. Carefully remove three nails from the windings. Connect the wire and deter-mine the number of paper clips the elec-tromagnet can hold. Record your results.

9. Determine whether the electromagnet at-tracts the penny, nickel, dime, and other test objects.

Observations

What materials are attracted to the electromagnet?

Analysis and Conclusions

1. What do the materials attracted to the magnet have in common?

2. When you increase the number of turns of wire, what effect does this have on the strength of the electromagnet?

3. How does removing the nails affect the strength of the electromagnet?

4. **On Your Own** How can you increase or decrease the strength of an electro-magnet without making any changes to the wire or to the nails? Devise an experi-ment to test your hypothesis.

Summarizing Key Concepts

3–1 Magnetism From Electricity

▲ A magnetic field is created around a wire that is conducting electric current.

▲ The relationship between electricity and magnetism is called electromagnetism.

▲ A coiled wire, known as a solenoid, acts as a magnet when current flows through it. A solenoid with a core of iron acts as a strong magnet called an electromagnet.

▲ A magnetic field exerts a force on a wire conducting current.

▲ An electric motor converts electric energy into mechanical energy that is used to do work.

▲ A galvanometer is a device consisting of an electromagnet attached to a needle that can be used to measure the strength and direction of small currents.

3–2 Electricity From Magnetism

▲ During electromagnetic induction, an electric current is induced in a wire exposed to a changing magnetic field.

▲ One of the most important uses of electromagnetic induction is in the operation of a generator, which converts mechanical energy into electric energy.

▲ A transformer is a device that increases or decreases the voltage of alternating current. A step-up transformer increases voltage. A step-down transformer decreases voltage.

Reviewing Key Terms

Define each term in a complete sentence.

3–1 Magnetism From Electricity
electromagnetism
solenoid
electromagnet
electric motor
galvanometer

3–2 Electricity From Magnetism
induced current
electromagnetic induction
generator
transformer

Chapter Review

Content Review

Multiple Choice

Choose the letter of the answer that best completes each statement.

1. The strength of the magnetic field of an electromagnet can be increased by
   a. increasing the number of coils in the wire only.
   b. increasing the amount of current in the wire only.
   c. increasing the amount of iron in the center only.
   d. all of these.

2. A generator can be considered the opposite of a(an)
   a. galvanometer.   c. electric motor.
   b. transformer.   d. electromagnet.

3. A device that changes the voltage of alternating current is a(an)
   a. transformer.   c. generator.
   b. electric motor.   d. galvanometer.

4. The scientist who discovered that an electric current creates a magnetic field is
   a. Faraday.   c. Henry.
   b. Oersted.   d. Maxwell.

5. A device with an electromagnet that continually rotates because of a changing electric current is a(an)
   a. doorbell.   c. galvanometer.
   b. solenoid.   d. electric motor.

6. The creation of an electric current by a changing magnetic field is known as
   a. electromagnetic induction.
   b. generation.
   c. transformation.
   d. stepping-up.

True or False

If the statement is true, write "true." If it is false, change the underlined word or words to make the statement true.

1. The relationship between electricity and magnetism is called electromagnetism.
2. A solenoid with a piece of iron in the center is called an electromagnet.
3. A commutator and brushes are found in an electric motor running on direct current.
4. A generator is used to detect small currents.
5. In a generator, mechanical energy is converted to electric energy.
6. Large generators at power plants often get their mechanical energy from steam.
7. An induced current is produced by a changing electric field.
8. A transformer changes the voltage of an electric current.

Concept Mapping

Complete the following concept map for Section 3–1. Refer to pages P6–P7 to construct a concept map for the entire chapter.

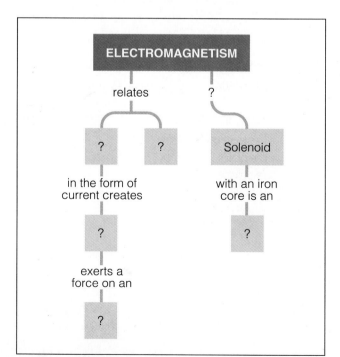

Concept Mastery

Discuss each of the following in a brief paragraph.

1. Does a magnetic field exert a force on a wire carrying electric current? Explain.
2. Describe how an electric motor operates. How does its operation differ depending on the type of current used to run the motor?
3. Explain how a galvanometer works.
4. Describe how a generator operates.
5. Describe the discoveries of Oersted and Faraday. How are these discoveries related?
6. What is a solenoid? An electromagnet?
7. What is a step-up transformer? A step-down transformer? Why are transformers important for the transmission of electricity?

Critical Thinking and Problem Solving

Use the skills you have developed in this chapter to answer each of the following.

1. **Making comparisons** Explain the difference between an electric motor and an electric generator in terms of energy conversion.
2. **Making diagrams** Use a diagram to show how the rotation of a wire loop in a generator first induces a current in one direction and then a current in the other direction.
3. **Applying concepts** The process of electromagnetic induction might seem to disobey the law of conservation of energy, which says that energy cannot be created. Explain why this is actually not so.
4. **Applying definitions** Indicate whether each of the following characteristics describes (a) a step-up transformer, (b) a step-down transformer, (c) both a step-up and a step-down transformer.
   a. Voltage in the secondary coil is greater.
   b. Involves electromagnetic induction.
   c. Voltage in the primary coil is greater.
   d. Used in doorbells and model trains.
   e. Consists of two insulated coils wrapped around opposite sides of an iron core.
   f. More loops in the secondary coil.
5. **Making inferences** Explain why a transformer will not operate on direct current.

6. **Making comparisons** Thomas Edison originally designed a power plant that produced direct current. Alternating current, however, has become standard for common use. Compare and contrast the two types of currents in terms of production and use.
7. **Using the writing process** Several devices that work on the principle of electromagnetism and electromagnetic induction have been mentioned in this chapter. Choose three of these devices and imagine what your life would be like without them. Write a short story or a poem that describes your imaginings.

Electronics and Computers

It was 1952—and not many people were familiar with computers. In fact, there were some who had never heard the word. But on Election Day of that year, millions of Americans came face to face with the computer age.

The presidential contest that year pitted Republican Dwight D. Eisenhower against Democrat Adlai E. Stevenson. Early in the evening, even before the voting polls had closed, newscaster Walter Cronkite announced to viewers than an "electronic brain" was going to predict the outcome of the election. The "electronic brain" was the huge UNIVAC I computer.

What UNIVAC predicted, based on just 3 million votes, was a landslide victory for Eisenhower. An amazed nation sat by their television sets into the early hours of the morning, convinced that the "electronic brain" could not have predicted as it did.

UNIVAC was not wrong, however. When all the votes were counted, the computer's prediction turned out to be remarkably close to the actual results. In this chapter you will learn about some of the devices that have brought the computer age and the electronic industry to where it is now.

Journal *Activity*

You and Your World   Do you watch television often? If so, what kinds of shows do you watch? If not, why not? Television is an electronic device in widespread use in modern society. In your journal, compare the positive contributions television has made with its negative aspects. Explain how you would improve the use of television.

◄ *A photographer's view of some of the electronic components that make computer technology possible.*

Guide for Reading

Focus on these questions as you read.

▶ *What is electronics?*

▶ *How are electrons related to electronic devices?*

4–1 Electronic Devices

Were you awakened this morning by the buzz of an alarm clock? Did you rely on a radio or a cassette player to get the day started with music? Did your breakfast include food that was warmed in a microwave oven? Can you get through the day without using the telephone or watching television?

You probably cannot answer these questions without realizing that electric devices have a profound effect on your life. The branch of technology that has developed electric devices is called **electronics.** Electronics is a branch of physics closely related to the science of electricity. **Electronics is the study of the release, behavior, and control of electrons as it relates to use in helpful devices.**

Although electronic technology is relatively new—it can be traced back only 100 years or so—electronics has rapidly changed people's lives. For example, telephones, radio, television, and compact disc players have revolutionized communication and

Figure 4–1 *Familiar electronic devices such as these have shaped the way people live and work. Have you used any of these devices?*

entertainment. Computers and robots have increased speed in business and industry. Electronic devices help physicians diagnose diseases and save lives. All forms of travel depend on electronic devices.

Perhaps you wonder how electronics differs from the study of electricity. After all, both deal with electrons and electric currents. The study of electricity concentrates on the use of electric currents to power a wide range of devices, such as lamps, heaters, welding arcs, and other electrical appliances. In such devices, the kinetic energy of moving electrons is converted into heat and light energy. Electronics treats electric currents as a means of carrying information. Currents that carry information are called electric signals. Carefully controlled, electrons can be made to carry messages, magnify weak signals, draw pictures, and even do arithmetic.

Vacuum Tubes

The ability to carefully control electrons began with the invention of the **vacuum tube.** The American inventor Thomas Edison (also known as the inventor of the phonograph) invented the first vacuum tube but unfortunately did not realize its importance. A simple vacuum tube consists of a filament, or wire, and at least two metal parts called electrodes. When a current flows through the filament, much as it does through a light bulb, the filament gives off heat. This heat warms the electrodes. One of the electrodes gives off electrons when heated. For this reason, it is called an emitter. This electrode is negatively charged. The other electrode, which does not give off electrons, is positively charged. This electrode is called the collector because electrons flow to it from the negatively charged emitter. As a result, a current of electrons flows through the vacuum tube. Thus a vacuum tube is a one-way valve, or gate, for a flow of electrons. Electrons are permitted to move in only one direction through the vacuum tube.

Both the electrodes and the filament are contained in a sealed glass tube from which almost all the air has been removed. How does this fact explain the name given to this tube? A vacuum tube may have several other parts between the electrodes. It

FIND OUT BY

DOING

Electronics in Your Home

Over the course of a day, observe and make a list of all the electronic appliances you use in your home.

■ Which of these appliances contain transistors? Cathode-ray tubes? Vacuum tubes? Integrated circuits?

■ Do some appliances contain multiple types of devices? If so, which ones?

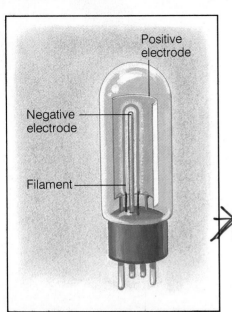

Positive electrode

Negative electrode

Filament

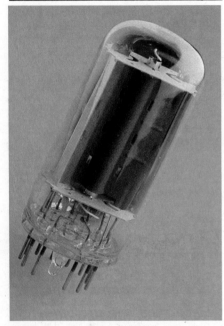

Figure 4–2 *A diode is the simplest type of vacuum tube. In a diode, electrons flow in one direction, from the negative electrode to the positive electrode. Diodes are used as rectifiers in many electronic devices. Why?*

may also have charged plates or magnets that can bend the stream of electrons.

Because vacuum tubes produce a one-way current, they have many applications in electronics. Two important applications of vacuum tubes are as rectifiers and as amplifiers. Vacuum tubes also acted as switches in early electronic computers.

Rectifiers

A **rectifier** is a vacuum tube that converts alternating current to direct current. The current supplied to your home is alternating current. Certain household appliances, however, cannot operate on alternating current. They need direct current. So these appliances have rectifiers built into their circuits. As the alternating current passes through the rectifier, it is changed into direct current.

The type of vacuum tube used most often as a rectifier is called a **diode.** A diode contains two electrodes. When alternating current is sent to a diode, the emitter will be charged by the current. However, it will emit a current only when it has a negative charge. And this happens only when the current is flowing in one direction. Thus the current leaving the diode is direct current.

Rectifiers are used in devices such as televisions, stereos, and computers. Converters that allow you to plug battery-operated devices into household electric outlets also contain rectifiers.

Amplifiers

An **amplifier** is an electronic device that increases the strength of an electric signal. (Remember, an electric signal is a current that carries information.) In an amplifier, a small input current is converted to a large output current. The large output current produces a stronger signal. The strengthening of a weak signal is called amplification. Amplification is perhaps the most important function of an electric device.

Another type of vacuum tube, a **triode,** is often used for amplification. A triode consists of a filament, a plate, and a wire screen, or grid. The addition of the grid allows the flow of electrons between the negatively charged emitter and the positively charged collector to be better controlled. The invention of the triode was responsible for the rapid growth of the radio and television industry.

Amplifiers strengthen both sound and picture signals. The signals that carry sound and picture information are often very weak as a result of traveling long distances through the air. By the time an antenna picks up the signals, they are too weak to produce an accurate copy of the original sound. Radio and television amplifiers strengthen the incoming signals. Fully amplified signals can be millions or billions of times stronger than the signal picked up by the antenna.

Without amplifiers, devices such as hearing aids, public-address systems, tape recorders, and radar would not operate. Amplifiers are also essential to the operation of medical instruments used to diagnose certain injuries and diseases. Human heart waves and brain waves can be studied by doctors because the weak electric signals given off by these organs are amplified.

Solid-State Devices

Electron devices can be divided into two main groups according to their physical structure. The vacuum tubes you have just read about make up one group. The other group consists of **solid-state devices.** Solid-state physics involves the study of the structure of solid materials.

From the 1920s until the 1950s, vacuum tubes dominated the world of electronics. In the 1950s, however, solid-state devices took over. The reason for this is obvious: Solid-state devices have several advantages over vacuum tubes. They are much smaller and lighter than vacuum tubes and give off much less heat. They also use far less electric power, are more dependable, and last longer. And in most cases, they are less expensive.

In solid-state devices, an electric signal flows through certain solid materials instead of through a vacuum. The use of solid-state devices was made possible by the discovery of **semiconductors.** Semiconductors are solid materials that are able

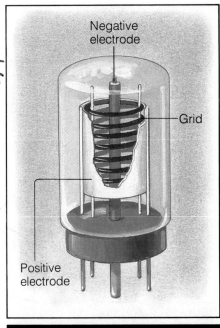

Figure 4–3 *A triode vacuum tube consists of a filament, a plate, and a grid. The addition of a grid allows the electron flow to be amplified, or strengthened. Triodes are used in microphones to amplify sound.*

Figure 4–4 *Devices that used vacuum tubes were large, heavy, and cumbersome. They also gave off a considerable amount of heat. Devices that use semiconductors can be made extremely small. In addition, they are more dependable, last longer, use less energy, and give off less heat.*

Figure 4–5 *Doping a semiconductor material increases its conductivity. In an n-type, the impurity adds extra electrons that can flow. In a p-type, doping creates holes that are missing electrons. What impurity is used to make an n-type semiconductor? A p-type?*

to conduct electric currents better than insulators do but not as well as true conductors do.

Silicon and germanium are the most commonly used semiconductors. These elements have structures that are determined by the fact that their atoms have four outermost electrons. Silicon and germanium acquire their usefulness in electronics when an impurity (atoms of another element) is added to their structure. Adding impurities to semiconductors increases their conductivity. The process of adding impurities is called **doping.**

There are two types of semiconductors. These types are based on the impurity used to dope the semiconductor. If the impurity is a material whose atoms have 5 outermost electrons (such as arsenic), the extra electrons will not fit into the structure of the semiconductor. Thus the doping contributes extra electrons that are somewhat free to move and that can form a current. This type of semiconductor is called an n-type, meaning negative-type. Silicon doped with arsenic is an n-type semiconductor.

If a semiconductor is doped with a material whose atoms have three outermost electrons (such as gallium), there will be empty holes in the semiconductor's crystal structure. These holes can also be used to form a current. Semiconductors doped with atoms that have fewer electrons, which is equivalent to saying extra protons, are called p-type semiconductors. Silicon doped with gallium is a p-type semiconductor. What do you think p-type means?

The arrangement of impurities on one type of semiconductor allows current to flow in only one

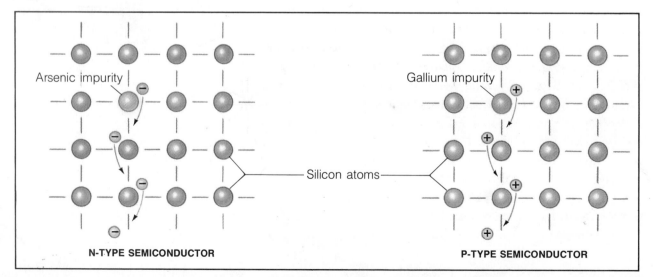

Arsenic impurity

Gallium impurity

Silicon atoms

N-TYPE SEMICONDUCTOR

P-TYPE SEMICONDUCTOR

direction. Thus this type of semiconductor acts as a diode. A different arrangement of impurities produces another solid-state device, which you are now going to read about.

Transistors

A **transistor** is a sandwich of three layers of semiconductors. A transistor is often used to amplify an electric current or signal. It is the arrangement of the impurities in the semiconductor that enables it to act as an amplifier. A weak signal, corresponding to a weak current, enters the transistor and is amplified so that a strong signal is produced.

Transistors come in a variety of shapes and sizes. Perhaps you are familiar with some of them. Transistors are commonly used in radios, televisions, stereos, computers, and calculators. The small size, light weight, and durability of transistors have helped in the development of communication satellites.

Figure 4-6 *Transistors come in a variety of shapes and sizes. What does a transistor do?*

Integrated Circuits

When you studied electric circuits in Chapter 1, you learned that a circuit can consist of many parts connected by wires. Complicated circuits organized in this fashion, however, become very large. In the 1960s, scientists found a way to place an entire circuit on a tiny board, thereby eliminating the need for separate components. This new type of circuit is known as an **integrated circuit.** An integrated circuit combines many diodes and transistors on a thin slice of silicon crystal. This razor-thin piece of silicon is often called a **chip.** A single integrated circuit, or chip, may often contain thousands of diodes and transistors in a variety of complex combinations.

To turn a silicon chip into an integrated circuit, it must be doped in some places with arsenic and in other places with gallium. Certain areas of the chip become diodes, while other areas become transistors. Connections between these diodes and transistors are then made by painting thin "wires" on the chip. Wires are attached to the integrated circuit so it can be connected to other devices.

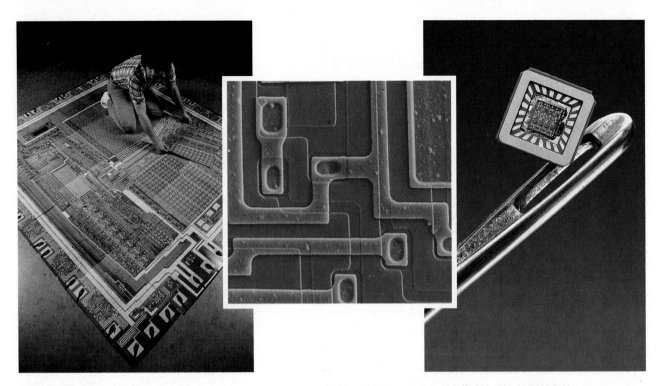

Figure 4–7 *An integrated circuit, or chip, contains thousands of diodes and transistors on a thin slice of silicon crystal (right). A computer scientist designs a new computer chip by first drawing a large version and then having the design miniaturized (left). Magnified 175 times by a scanning electron microscope, the integrated circuit paths can be seen (center). A human hair is wider than 150 of these paths!*

Integrated circuits are used as amplifiers and switches in a wide variety of devices. Computers and microcomputers, calculators, radios, watches, washing machines, refrigerators, and even robots use integrated circuits.

4–1 Section Review

1. What is electronics? How is it different from the study of electricity?
2. How are electrons used in a vacuum tube? How is a vacuum tube used as a rectifier? As an amplifier?
3. What is a semiconductor?
4. How are semiconductors used to make integrated circuits? What are some advantages of the use of integrated circuits?

Connection—*Language Arts*
5. The words rectify and amplify are not limited to scientific use. Explain what these words mean and give examples of their use in everyday language. Then explain why they are appropriate for the electronic devices they name.

4-2 Transmitting Sound

Since the first telegraph line was connected and the first telegraph message was sent in 1844, people have become accustomed to instant communication. Each improvement in the speed, clarity, and reliability of a communication device has been based on a discovery in the field of electronics.

One particular discovery that is the basis for many devices used to transmit information is an interesting relationship between electricity and magnetism. The discoveries made by Oersted, Faraday, and others that you have read about in Chapter 3 clearly illustrate that electricity and magnetism are related. Another scientist, James Clerk Maxwell, used the discoveries of his predecessors to open a new world of scientific technology.

Maxwell showed that all electric and magnetic phenomena could be described by using only four equations involving electric and magnetic fields. Thus he unified in one theory all phenomena of electricity and magnetism. These four equations are as important to electromagnetism as Newton's three laws are to motion.

Perhaps the most important outcome of Maxwell's work is the understanding that not only does a changing magnetic field give rise to an electric field, but a changing electric field produces a magnetic field. In other words, if a magnetic field in space is changing, like the up and down movements of a wave, a changing electric field will form. But a changing electric field will also produce a changing magnetic field. The two will keep producing each other over and over. The result will be a wave consisting of an electric field and a magnetic field. Such a wave is called an **electromagnetic wave.** Electromagnetic waves are like waves on a rope except that they do not consist of matter. They consist of fields.

You are already more familiar with electromagnetic waves than you may realize. The most familiar electromagnetic wave is light. All the light that you see is composed of electromagnetic waves. The microwaves that heat food in a microwave oven are also electromagnetic waves. The X-rays that a doctor takes are electromagnetic waves. And as you will

Guide for Reading

Focus on this question as you read.

▶ *How do sound-transmitting devices work?*

FIND OUT BY WRITING

Telephone and Radio History

The invention of the telephone and the invention of the radio were two important advances in electronic technology. Using books and other reference materials in the library, find out about the invention of each. Be sure to include answers to the following questions.

1. Who invented the device?

2. When was it invented?

3. Were there any interesting or unusual circumstances surrounding the invention?

4. How was the invention modified through the years?

Present the results of your research in a written or oral report. Accompany your report with illustrations.

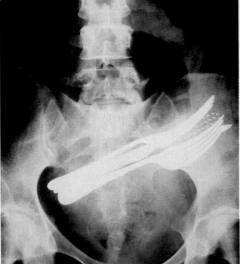

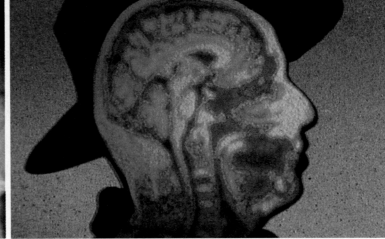

Figure 4–8 *Electromagnetic waves in the form of X-rays enable you to see the forks, pen, and toothbrush in this person's intestine! Another type of EM wave allows scientists to study the composition of the brain and other parts of the body. Believe it or not, this photograph of a river was taken in total darkness using infrared waves. What type of electromagnetic wave can be seen in the bottom photograph?*

soon learn, the prediction and verification of electromagnetic waves opened up a whole new world of communication—from the first wireless telegraph to radio and television to artificial space satellites.

Radio Communication

Would you be surprised to learn that radio broadcasting once played the same role that television plays today? People would gather around a radio to listen to a variety of programs: musical performances, comedy shows, mystery hours, and news broadcasts. Today, radio broadcasting is still a great source of entertainment and information. But its role has been greatly expanded.

Radios work by changing sound vibrations into electromagnetic waves called radio waves. The radio waves, which travel through the air at the speed of light, are converted back into sound vibrations when they reach a radio receiver.

A radio program usually begins at a radio station. Here, a microphone picks up the sounds that are being broadcast. An electric current running through the microphone is disturbed by the sound vibrations in such a way that it creates its own vibrations that match the sound.

The electric signals that represent the sounds of a broadcast are now sent to a transmitter. The transmitter amplifies the signals and combines them with a radio wave that will be used to carry the information. How the wave carries the information determines if the information will be broadcast as AM or FM radio. The final wave carrying the signal is then sent to a transmitting antenna. The antenna sends

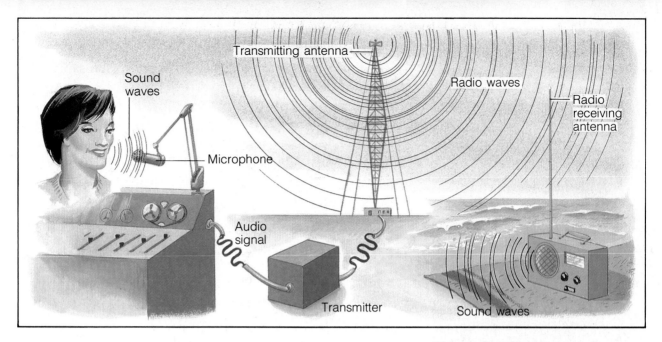

Figure 4–9 *Radios work by converting sound vibrations into electromagnetic waves. The waves are amplified and sent out into the air. Picked up by a receiving antenna, the radio waves are converted back into sound waves. What else does the receiver do with the radio waves?*

the radio waves out into the air. Why do you think many radio stations locate their antennas at high elevations, in open areas, or on top of towers?

Radio waves are converted back into sound waves by means of a radio receiver. A radio receiver picks up and amplifies the radio waves originally sent out from a radio station. When a radio receiver picks up sounds corresponding to a specific frequency, it is described as being tuned in.

Have you ever wondered how a telephone call can be made from a moving vehicle such as a car or an airplane? Cellular telephones (movable telephones) use signals that are sent out by an antenna as radio waves. Other antennas pick up the radio waves and convert them back into sound. To understand more about cellular telephones, you must first learn how a telephone operates.

Telephone Communication

Have you ever stopped to think about the amazing technology that enables you to talk to someone else, almost anywhere in the world, in seconds? The device that makes this communication possible is the telephone. The first telephone was invented in 1876 by Alexander Graham Bell. Although modern telephones hardly resemble Bell's, the principle on which all telephones work is the same. A telephone sends and receives sound by means of electric signals. A telephone has two main parts.

Figure 4–10 *The first telephone was invented in 1876 by Alexander Graham Bell. Today, you push buttons or dial to make calls. But up until the early 1950s, telephone calls were placed by switchboard operators, whose familiar phrase was "Number, please."*

TRANSMITTER The transmitter is located behind the mouthpiece of a telephone. Sound waves produced when a person speaks into a telephone cause a thin metal disk located in the transmitter to vibrate. These vibrations vary according to the particular sounds. The vibrations, in turn, are converted to an electric current. The pattern of vibrations regulates the amount of electric current produced and sent out over telephone wires. You can think of the electric current as "copying" the pattern of the sound waves. The electric current travels over wires to a receiver.

RECEIVER The receiver, located in the earpiece, converts the changes in the amount of electric current sent out by a transmitter back into sound. The receiver uses an (electromagnet) to produce this conversion.

When the electric current transmitted by another telephone goes through the coil of the electromagnet, the electromagnet becomes magnetized. It pulls on another thin metal disk, causing it to vibrate. The vibrations produce sound waves that the listener hears.

Alexander Graham Bell used carbon grains in the transmitter to convert sound to electricity. Today's telephones use a small semiconductor crystal. Transistors then amplify the electric signal. In modern telephone earpieces, semiconductor devices have replaced electromagnets. And fiber-optic systems, in which the vibrations travel as a pattern of changes in a beam of laser light, are replacing copper cables used for ordinary transmission.

Figure 4–11 *A telephone, consisting of a transmitter and a receiver, sends and receives sounds by means of electric signals. Where are transistors used? For what purpose?*

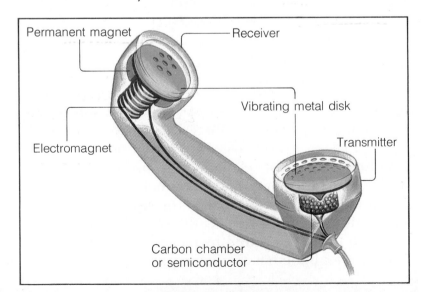

4-2 Section Review

1. Describe the relationship between sound and electric current in devices that transmit sound.
2. Describe the two main parts of a telephone.
3. Describe the broadcast of a radio program.
4. How do you think solid-state devices have affected telephones and radios?

Connection—*You and Your World*
5. How do you think radio communication has affected the development of business and industry?

4-3 Transmitting Pictures

You would probably agree that a video game would be far less exciting if the images were unclear and did not move very quickly. The same is true of your favorite television show. The images on a video screen and a television screen are produced by a special type of vacuum tube.

Cathode-ray Tubes

Television images are produced on the surface of a type of vacuum tube called a **cathode-ray tube,** or CRT. Cathode-ray tubes are also responsible for images produced by video games, computer displays, and radar devices.

A cathode-ray tube is an electronic device that uses electrons to produce images on a screen. This special type of vacuum tube gets its name from the fact that inside the glass tube, a beam of electrons (cathode rays) is directed to a screen to produce a picture. The electrons, moving as a beam, sweep across the screen and cause it to glow. The screen glows because it is coated with fluorescent material. Fluorescent material glows briefly when struck by electrons.

The electrons in a CRT come from the negatively charged filament within the sealed glass vacuum tube. An electric current heats the metal filament

Guide for Reading

Focus on this question as you read.

▶ *How does a cathode-ray tube operate?*

Figure 4-12 *When you play a video game, you are taking advantage of a cathode-ray tube.*

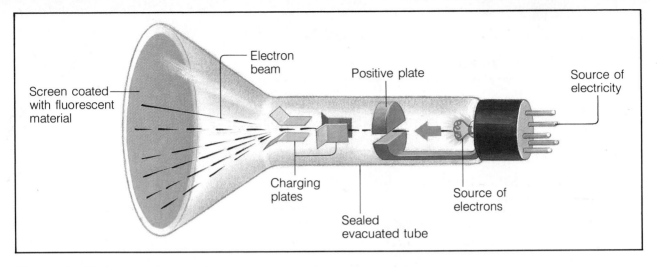

Figure 4–13 *A cathode-ray tube is a sealed evacuated tube in which a beam of electrons is focused on a screen coated with fluorescent material. As electrons strike the fluorescent material, visible light is given off and an image is formed.*

and causes electrons to "boil" off it. The electrons are accelerated toward the screen and focused into a narrow beam. Because the electrons move so quickly in a concentrated beam, the source is sometimes referred to as an electron gun. The moving electrons produce a magnetic field that can be used to control the direction of the beam. Electromagnets placed outside the CRT cause the beam to change its direction, making it move rapidly up and down and back and forth across the screen.

At each point where the beam of electrons strikes the fluorescent material of the CRT screen, visible light is given off. The brightness of the light is determined by the number of electrons that strike the screen. The more electrons, the brighter the light. The continuous, rapid movement of the beam horizontally and vertically across the screen many times per second produces a pattern of light, or a picture on the screen. In the United States, the electron beam in a CRT traces 525 lines as it zigzags up and down, creating a whole picture 30 times each second. In some other countries the beam moves twice as fast, creating an even clearer image.

Television Transmission

A cathode-ray tube in a color television set differs from a simple cathode-ray tube in two important ways. First, the screen of a color television set is coated with three different materials placed close together in clusters of dots or in thin stripes at each point on the screen. Each material glows with a different color of light—red, blue, or green—when struck by a beam of electrons. Various colors are

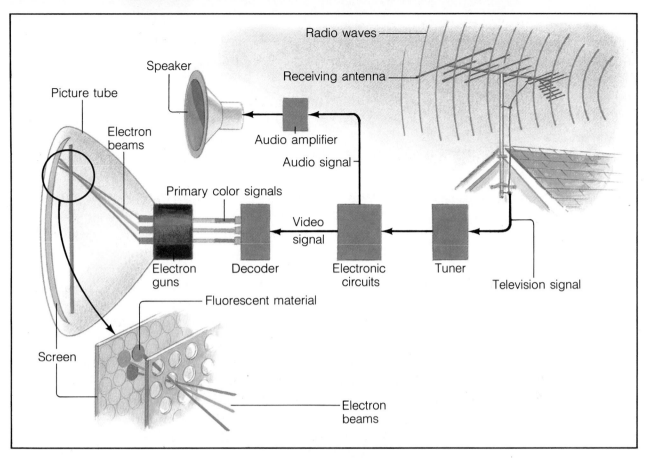

Radio waves

Speaker

Receiving antenna

Picture tube

Electron beams

Audio amplifier

Audio signal

Primary color signals

Video signal

Electron guns

Decoder

Electronic circuits

Tuner

Television signal

Fluorescent material

Screen

Electron beams

produced by adjusting the strengths of the electron beams. For example, red is produced when electrons strike only the red material. Purple is produced when electrons strike both red and blue material. When do you think white is produced?

Second, a color television CRT contains three electron guns—one for each color (red, green, and blue). The information for controlling and directing the beams from the three electron guns is coded within the color picture signal that is transmitted from a TV station.

Figure 4–14 *The CRT in a color television contains three electron guns—one each for red, blue, and green signals. The screen of the CRT is coated with three different fluorescent materials, each of which glows with a different primary color of light when struck by a beam of electrons.*

4–3 Section Review

1. What is a cathode-ray tube? Describe how it works.
2. How does a color television CRT differ from a simple CRT?

Critical Thinking—*Making Inferences*
3. Photographs of a TV picture taken with an ordinary camera often show only part of the screen filled. Explain why.

4–4 Computers

Computers have quickly become a common sight over the past few decades. You see computers in stores, doctors' and dentists' offices, schools, and businesses. Perhaps you even have one in your home. **A computer is an electronic device that performs calculations and processes and stores information.** A modern electronic computer can do thousands of calculations per second. At equally incredible speed, it can file away billions of bits of information in its memory. Then it can rapidly search through all that information to pick out particular items. It can change numbers to letters to pictures to sounds—and back to numbers again.

Using these abilities, modern computers are guiding spaceships, navigating boats, diagnosing diseases and prescribing treatment, forecasting weather, and searching for ore. Computers make robots move, talk, and obey commands. Computers can play games and make music. They can even design new computers. The pages of this textbook were composed and printed with the help of a computer (although people still do the writing)!

Computer Development

The starting point of modern computer development is thought to be 1890. In preparation for the United States census that year, Herman Hollerith

Figure 4–15 *Early computers, which used large vacuum tubes were neither fast nor reliable. And like the ENIAC, they certainly were enormous in size! A modern computer that fits on a desk top once required an entire room.*

devised an electromagnetic machine that could handle information punched into cards. The holes allowed small electric currents to pass through and activate counters. Using this system, Hollerith completed the 1890 census in one fourth the time it had taken to do the 1880 census! Hollerith's punch card became the symbol of the computer age.

The first American-built computer was developed in 1946 by the United States Army. The Electronic Numerical Integrator and Calculator, or ENIAC, consisted of thousands of vacuum tubes and occupied a warehouse. It cost millions of dollars to build and maintain. It was constantly breaking down and had to be rebuilt each time a new type of calculation was done. ENIAC required great amounts of energy, generated huge amounts of heat, and was very expensive. By today's standards, ENIAC was slow. It could do only 100,000 calculations per second.

The first general-purpose computer was introduced in 1951. It was called the Universal Automatic Computer, or UNIVAC. UNIVAC was certainly an improvement over ENIAC, but it was still large, expensive, and slow.

Increased demand for computers encouraged more advanced computer technology. Technical breakthroughs such as transistors and integrated

FIND OUT BY
CALCULATING

Computing Speed

Shuffle a deck of playing cards. Have a friend time you as you sort the cards, first into the four suits, and then from the 2 through the ace in each suit. Determine how many sorts you made.

Calculate how many sorts you made per second.

A bank check-sorting machine can make 1800 sorts per minute.

■ How much faster than you is this machine?

Figure 4–16 *The uses of computers are wide and varied. Computer applications include the identification of worldwide ozone concentrations (bottom left), the analysis of the body mechanics of a runner (top), and the study of fractal geometry in mathematics (bottom right).*

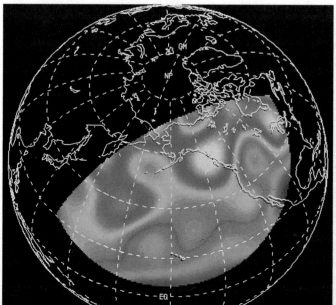

circuits reduced the size and cost of computers. They also increased the efficiency, speed, and uses of computers. And equally important, they brought the computer within everyone's reach.

The future of computers lies in both the very small and the very large. Integrated circuits called microprocessors can hold an entire processing capability on one small chip. At the other extreme, groups of computers are being linked together to form supercomputers.

Computer Hardware

Computer **hardware** refers to the physical parts of a computer. **Computer hardware includes a central processing unit, main storage, input devices, and output devices.**

The "brain" of a computer is known as the **central processing unit,** or CPU. A CPU controls the operation of all the components of a computer. It executes the arithmetic and logic instructions that it receives in the form of a computer program. A computer program is a series of instructions that tells the computer how to perform a certain task. A computer program can be written in one of several different computer languages.

The main storage of a computer is often referred to as the **main memory.** The main memory contains data and operating instructions that are processed by the CPU. In the earliest computers, the main memory consisted of thousands of vacuum tubes. Modern computer memory is contained on chips. The most advanced memory chip can store as much information as 1 million vacuum tubes can.

Data are fed to the central processing unit by an **input device.** One common input device is a keyboard. A keyboard looks very much like a typewriter. Using a keyboard, a person can communicate data and instructions to a computer. Other input devices include magnetic tape, optical scanners, and disk drives.

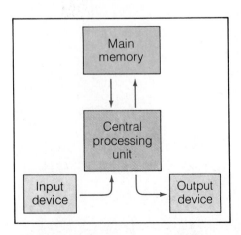

Figure 4–17 *Computer hardware includes a central processing unit, main memory, an input device, and an output device. What is the function of each?*

A **disk drive** reads information off a disk and enters it into the computer's memory or into the CPU. Information from a disk drive can be placed into a computer very quickly.

Information produced by a computer can be removed and stored on a disk. So a disk drive is also an **output device.** An output device receives data from the central processing unit. Output devices include printers, cathode-ray tubes, magnetic tape drives, and voice synthesizers. Even robots are output devices.

CONNECTIONS

The Human Computer

The organization of the human body is pretty amazing, isn't it? Think of all the various organs constantly working independently as well as interacting with one another to keep you alive, healthy, and functioning normally. What's even more amazing is that the body does all its work without your having to think about it! You don't even have to worry about nourishing your body because it reminds you to do so by making you feel hungry. Now that's pretty awesome! What is even more incredible, perhaps, is that you have the ability to think, reason, reach conclusions, and use your imagination. What a wonderful device the human body is!

The organ of your body that controls the various body systems and is also the seat of intelligence is the *brain.* Different parts of the brain have different functions. One part enables you to coordinate your movements quickly and gracefully. Another part controls body processes, such as heartbeat, breathing, and blood pressure—not to mention swallowing, sneezing, coughing, and blinking. And still another part is responsible for your ability to think.

Many attempts have been made to simulate the activities of the brain. There is, in fact, a field of computer research that aims to create artificial intelligence—that is, computers capable of thinking and reasoning like humans. Currently, however, such attempts to unravel the interconnections of the human brain have proved largely unsuccessful. For now, at least, the human brain still holds its many secrets.

Like a disk drive, a **modem** is an input and output device. A modem changes electronic signals from a computer into signals that can be carried over telephone lines. It also changes the sounds back into computer signals. A modem allows a computer to communicate with other computers, often thousands of kilometers away. As computers link in this way, they form a network in which information can be shared. A modem allows use of this network by accessing (getting) information from a central data bank. A data bank is a vast collection of information stored in a large computer.

The Binary System

Computer hardware would be useless if computer **software** did not exist. **Software is the program or set of programs the computer follows.** Software must be precise because in most cases a computer cannot think on its own. A computer can only follow instructions. For example, to add two numbers, a program must tell a computer to get one number from memory, hold it, get the other number from memory, combine the two numbers, and print the answer. After completing the instruction, the computer must be told what to do next.

A computer can execute instructions by counting with just two numbers at a time. The numbers are 0 and 1. The system that uses just these two numbers is called the **binary system.**

Computer circuits are composed of diodes. As you learned in the previous sections, diodes are gates that are either open or closed to electric current. If the gate is open, current is off. If the gate is closed, current is on. To a computer, 0 is current off and 1 is current on. Each digit, then, acts as a tiny electronic switch, flipping on and off at unbelievable speed.

Each single electronic switch is called a **bit.** A string of bits—usually 8—is called a **byte.** Numbers, letters, and other symbols can be represented as a byte. For example, the letter A is 01000001. The letter K is 11010010. The number 9 is 00001001.

You do not need to be reminded of the importance of computers. You have only to look around you. The uses of computers are many, and their presence is almost universal. Any list of computer

PROBLEM Solving

Go With the Flow

A computer follows a series of activities that take place in a definite order, or process. If you think about it, so do you. You get dressed one step at a time. When you follow a recipe, you add each ingredient in order.

It is useful to have a method to describe such processes, especially when writing computer programs. A flowchart is one method of describing a process. In a flowchart the activities are written within blocks whose shapes indicate what is involved in that step.

Designing a Flowchart It is a good thing you know how to write a flowchart because you are hosting the class luncheon today. But the chef that you have hired can cook only from a flowchart. Write a flowchart showing the chef how to prepare today's menu.

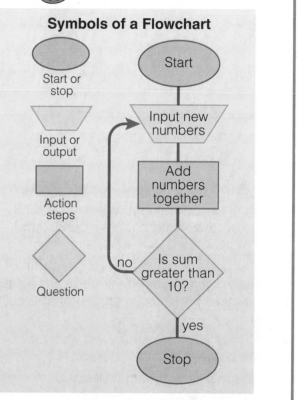

applications cannot really be completed today. For by the time today is over, another application will have been devised. The future of computers is exciting, indeed!

4–4 Section Review

1. What is a computer? How are its hardware and software involved in its operation?
2. What is a modem? How is it related to a data bank?
3. How is the binary system used by a computer?

Critical Thinking—*Making Calculations*
4. Show how the following numbers would be represented by a byte: 175, 139, 3, 45, 17.

FIND OUT BY READING

Artificial Intelligence

It is difficult to imagine using wires and crystals to construct a device that can think as you can. Read David Gerrold's *When H.A.R.L.I.E. Was One* and discover what such a device would be like.

Laboratory Investigation

The First Calculator: The Abacus

Problem

How can an abacus serve as a counting machine?

Materials *(per group)*

abacus

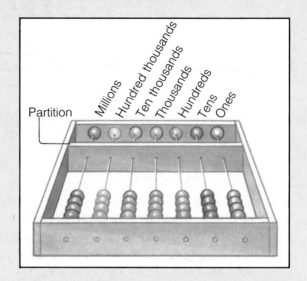

Procedure

1. The columns of beads on the abacus represent, from right to left, units of ones, tens, hundreds, thousands, and millions.

2. The single bead in the upper section of each column, above the partition, equals 5 beads in the lower section of that column.

3. Always start from the lower section of the far right, or ones, column.

4. Count to 3 by sliding 3 beads of the ones column to the partition.

5. Continue counting to 8. Slide the fourth bead up to the partition. You should be out of beads in this section. Slide all 4 beads back down and slide the single bead from the upper section of this column down to the partition. Remember that the top bead equals 5 lower beads. Continue counting from 6 to 8 by sliding the beads in the lower section of the ones column up to the partition. Before doing any further counting, check with your teacher to see that you are using the abacus correctly.

6. Continue counting to 12. Slide the last bead in the ones column up to the partition. You should now have a total of 9. And you should be out of beads in the ones column. Slide all these beads back

to their original zero position. Slide 1 bead in the lower section of the tens column up to the partition. This represents 10. Continue counting in the ones column until you reach 12.

Observations

1. Count to each of these numbers on the abacus: 16, 287, 5016, 1,816,215.

2. How would you find 8 + 7 on the abacus? Start by counting to 8 on the abacus. Then continue adding 7 more beads Find the following sums: 7 + 8, 3 + 4, 125 + 58.

Analysis and Conclusions

1. On what number system is the operation of the abacus based?

2. How does this compare with the operation of a computer?

3. **On Your Own**   Try designing a number system that uses more than 2 numerals but fewer than 10. Count to 20 in your system.

Study Guide

Summarizing Key Concepts

4–1 Electronic Devices

▲ Electronics is the study of the release, behavior, and control of electrons as it relates to use in practical devices.

▲ In a tube in which most of the gases have been removed—a vacuum tube—electrons flow in one direction.

▲ A rectifier is an electronic device that converts alternating current to direct current.

▲ An amplifier is an electronic device that increases the strength of an electric signal.

▲ Semiconductors are materials that are able to conduct electric currents better than insulators do but not as well as true conductors do.

▲ Adding impurities to semiconductors is called doping.

▲ An integrated circuit combines diodes and transistors on a thin slice of silicon crystal.

4–2 Transmitting Sound

▲ Radios work by changing sound vibrations into electromagnetic waves, or radio waves.

▲ A telephone sends and receives sound by means of electric signals. A telephone has two main parts: a transmitter and a receiver.

4–3 Transmitting Pictures

▲ A cathode-ray tube, or CRT, is an electronic device that uses electrons to produce images on a screen.

▲ A cathode-ray tube in a color television set contains three electron guns—one each for red, blue, and green signals.

4–4 Computers

▲ Integrated circuits called microprocessors can hold the entire processing capability on one small chip.

▲ Computer hardware consists of a central processing unit, main storage, input devices, and output devices.

▲ A computer program is a series of instructions that tell the computer how to perform a certain task.

Reviewing Key Terms

Define each term in a complete sentence.

4–1 Electronic Devices
electronics
vacuum tube
rectifier
diode
amplifier
triode
solid-state device
semiconductor
doping
transistor
integrated circuit
chip

4–2 Transmitting Sound
electromagnetic wave

4–3 Transmitting Pictures
cathode-ray tube

4–4 Computers
hardware
central processing unit
main memory
input device
disk drive
output device
modem
software
binary system
bit
byte

Chapter Review

Content Review

Multiple Choice

Choose the letter of the answer that best completes each statement.

1. Diode vacuum tubes are used as
   a. transistors.　　　c. amplifiers.
   b. rectifiers.　　　d. triodes.
2. Which electronic device was particularly important to the rapid growth of the radio and television industry?
   a. triode　　　c. punch card
   b. diode　　　d. disk drive
3. Which of the following is a semiconductor material?
   a. copper　　　c. silicon
   b. plastic　　　d. oxygen
4. A sandwich of three semiconductor materials used to amplify an electric signal is a(an)
   a. diode.　　　c. integrated circuit.
   b. transistor.　　　d. modem.
5. Radios work by changing sound vibrations into
   a. cathode rays.　　　c. electric signals.
   b. gamma rays.　　　d. bytes.
6. Which is not an advantage of solid-state devices in telephones and radios?
   a. smaller size
   b. increased cost
   c. better amplification
   d. greater energy efficiency
7. The physical parts of a computer are collectively referred to as computer
   a. software.　　　c. programs.
   b. peripherals.　　　d. hardware.
8. Which is computer software?
   a. printer　　　c. program
   b. disk drive　　　d. memory

True or False

If the statement is true, write "true." If it is false, change the underlined word or words to make the statement true.

1. A device that converts alternating current to direct current is a <u>rectifier</u>.
2. A telephone sends and receives sound by means of <u>electric</u> current.
3. The beam of <u>electrons</u> in a <u>cathode-ray tube</u> produces a picture.
4. A color television CRT has <u>two</u> electron guns.
5. The first American-built computer was <u>UNIVAC</u>.
6. <u>Microprocessors</u> are integrated circuits that can hold the entire processing capability of a computer on one chip.
7. <u>Output</u> devices feed data to a computer.
8. A <u>data bank</u> is a vast collection of information stored in a large computer.
9. A string of bits, usually eight in number, is called a <u>byte</u>.

Concept Mapping

Complete the following concept map for Section 4–1. Refer to pages P6–P7 to construct a concept map for the entire chapter.

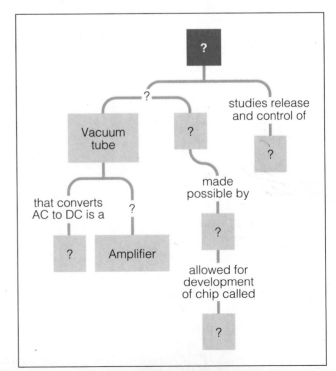

Concept Mastery

Discuss each of the following in a brief paragraph.

1. Why are electrons important to electronic devices? Give some examples.
2. Compare the functions of rectifiers and amplifiers. What type of vacuum tube is used for each?
3. Describe a semiconductor. Why are semiconductors doped?
4. What connection between electricity and magnetism is used in radio communication systems?
5. How does a telephone work?
6. How is a radio show broadcast?
7. In what ways are semiconductor diodes better than their vacuum tube ancestors?
8. Describe how a cathode-ray tube creates a picture.
9. What is a computer? What are some uses of computers?
10. What system is used in computers and calculators? Explain how it works.

Critical Thinking and Problem Solving

Use the skills you have developed in this chapter to answer each of the following.

1. **Making diagrams** Draw a diagram that shows how the four main hardware components of a computer are related.
2. **Sequencing events** The sentences below describe some of the energy conversions required for a local telephone call. Arrange them in proper order.
   a. Sound vibrates a metal plate.
   b. An electromagnet is energized.
   c. A vibrating magnet produces sound.
   d. Vibrating vocal cords produce sound.
   e. Mechanical energy is converted into an electric signal.
3. **Classifying computer devices** Many methods for putting data into a computer are similar to methods for getting data out of a computer. Identify each of the following as an input device, an output device, or both: typewriter keyboard, CRT, printer, optical scanner, magnetic tape, disk drive, punched cards, voice synthesizer.
4. **Applying concepts** A program is a list of instructions that tells a computer how to perform a task. Write a program that describes the steps involved in your task of waking up and arriving at school for your first class.
5. **Making calculations** The pictures on a television screen last for one thirtieth of a second. How many pictures are flashed on a screen during 30 minutes?
6. **Using the writing process** The electronics industry and the development of computers have evolved quite rapidly. You yourself are even witnessing obvious changes in the size and capabilities of electronic devices such as calculators, video games, computers, VCRs, compact disc players, and videodiscs. Write a story or play about how you think electronic devices will affect your life thirty or forty years from now.

GAZETTE
THE SEARCH FOR
SUPERCONDUCTORS

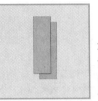

magine trains that fly above their tracks at airplane speeds and powerful computers that fit in the palm of your hand. Picture unlocking the secrets of the atom, or skiing on slopes made of air. Purely imagination? Not really. All of these things—and more—have been brought closer to reality by the work of Dr. Karl Alex Mueller and Dr. Johannes Georg Bednorz. These two dedicated scientists have changed fantasy to fact through their work with superconductors.

SEARCHING FOR A BETTER CONDUCTOR

Much of our electricity runs through copper wire. Copper is an example of a conductor, or a material that carries electricity well. However, copper is not a perfect con-

ductor because it offers resistance to the flow of electricity. As a result of resistance, about 15 percent of the electric power passing through a copper wire is wasted as heat.

A superconductor has no resistance. Therefore, it can conduct electricity without any loss of power. With superconductors, power plants could produce more usable electricity at lower costs and with no waste. Electric motors could be made smaller and more powerful. Superconducting wires connecting computer chips could produce smaller, faster computers.

Scientists have known about superconductors for more than 75 years. But although the principle of superconductivity was understood, the method of creating one remained a secret...a secret that seemed to be "locked in a deep freeze." For until the time of Mueller and Bednorz's discovery, materials would not become superconductors unless they were chilled to at least –250°C!

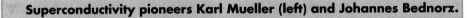

Superconductivity pioneers Karl Mueller (left) and Johannes Bednorz.

To cool materials to such extremely low temperatures, scientists had to use liquid helium, which is very costly. The supercold superconductors were just too expensive to be of general use.

If a substance could become superconducting at −196°C or higher, then it could be cooled with liquid nitrogen. Liquid nitrogen costs as little as a nickel a liter—less expensive than milk or soda. But what substances might become superconductors at these relatively high temperatures? That was the problem Dr. Mueller and Dr. Bednorz had to solve.

LOOKING IN A NEW DIRECTION

Many experts thought that superconductors simply did not exist at temperatures higher than −250°C. But Dr. Mueller, a highly respected physicist at IBM's research laboratory in Zurich, Switzerland, remained fascinated by high-temperature superconductors. In fact, he had already devised a new approach to finding one!

To some, his idea seemed impossible. But Dr. Mueller and his partner Dr. Bednorz were willing to follow their unusual approach under the guidance of what Dr. Mueller describes as "my intuition."

For almost three years, the two scientists mixed powders, baked them in ovens to form new compounds, and then chilled them to see if they would lose their resistance to electricity. And for three years, the two scientists kept their work a secret. "We were sure anybody would say, 'These guys are crazy,'" Dr. Bednorz later said. But despite endless hours of hard work and dedication, none of the new compounds was the superconductor Mueller and Bednorz sought.

Then in December 1985, Dr. Bednorz read about a new copper oxide. He and Dr. Mueller thought the oxide looked promising. They decided to test it for superconductivity. On January 27, 1986, Dr. Mueller and Dr. Bednorz broke the temperature barrier to superconductivity—and broke it by a

▲ This magnetic disk may seem to be defying gravity. Actually, it is floating above a disk made of a superconductive material. The superconductive disk repels magnetic fields and causes the magnet to float in midair.

large amount. They achieved superconductivity at −243°C. By April, Mueller and Bednorz had raised the temperature of superconductivity to a new record, −238°C. Around the world, scientists began to duplicate the experiments and make even greater advances in high-temperature superconductors.

In February 1987, a team of researchers at the University of Houston led by Dr. C.W. Chu created a new oxide that shows superconductivity at −175°C. This is the first superconductor that can be cooled with liquid nitrogen—the first superconductor that might be used for everyday purposes.

Dr. Mueller and Dr. Bednorz received the 1987 Nobel Prize for Physics for their pioneering work on superconductors. Their work, however, does not end here. They look forward to the development of a room-temperature superconductor!

ELECTRICITY: CURE-ALL OR END-ALL?

You have probably read or heard the story about Benjamin Franklin's discovery of electricity during a lightning storm. In the midst of a downpour, he and his son flew a kite with a key attached to the string. The shock they received by touching the key proved that the lightning's charge was conducted through the string and into the metal key. But could Franklin have imagined the impact of his discovery? The modern world of electric appliances, computers, stereos, washing machines, air conditioners, heaters, and more has been built upon his insight. In fact, much of society has come to rely on electricity for most of its needs. Electricity seems like the perfect symbol of technology—the answer to every modern need. Or is it?

In recent years, an increasing number of scientific reports suggest that electric current may have harmful biological effects. It is not the danger of the electric current itself that has people worried. Rather, it is the electromagnetic fields produced by certain levels of electric current in sources such as power lines, household wiring, and electric appliances. Such emissions are referred to as electromagnetic radiation. In particular, there are suggestions that there is a link between exposure to these electromagnetic fields and incidents of cancer.

A connection between high levels of electromagnetic radiation and harmful biological effects has been recognized for quite some time. But long-term exposure to levels of such high intensity are rare. Concern over the dangers caused by low levels of electromagnetic radiation, which began in the late 1970s, created new fears. Virtually all industrialized populations are commonly exposed to such levels of electromagnetic radiation.

The subject came under close scrutiny when Nancy Wertheimer, an epidemiologist (scientist who studies diseases), set out to study the homes of children who had died from cancer in a Colorado neighborhood in order to establish some connection. To her surprise, she found a statistical connection between childhood cancers and exposure to high-current power lines.

Wertheimer's work heralded a series of investigations and a bitter controversy. A similar study in the Colorado area supported her general conclusions, as did another study conducted in Europe. Several other research projects went on to show that workers regularly exposed to strong electromagnetic fields—electricians, power-station operators, telephone line workers—developed and died from leukemia, brain cancer, and brain tumors at significantly higher rates than workers in other fields. These studies also suggested a link between exposure to electromagnetic fields and low-grade illnesses such as fatigue, headaches, drowsiness, and nausea. Additional studies showed a link between the use of electric blankets and miscarriages (lost pregnancies).

▲ The electromagnetic fields in the area surrounding these high-voltage power lines are strong enough to light the fluorescent bulbs.

▼ Prolonged exposure to electromagnetic fields, a hazard of certain jobs, may be related to a high incidence of brain tumors. The gamma image shows a brain with a tumor.

Some of these studies and their related legal battles have prompted electric companies to reroute their wiring and change the location of their transformers. Often, however, cases regarding electromagnetic radiation hazards have been dismissed or postponed due to lack of clear evidence. Despite the seriousness of the issues, the research found to date does not prove a clear, direct link between electromagnetic radiation and cancer. For this reason such findings have not convinced everyone in the scientific and medical communities, the nation's courts, or those in industry of the possible dangers of electromagnetic fields. Skeptical researchers and judges complain that the studies lack scientific foundations, and that they show only a statistical link. Critics say that they need to see actual proof of a cause-and-effect relationship between the electromagnetic fields and illness.

To meet the skeptics' challenges, scientists are trying to show how electromagnetic fields actually affect and harm human cells. Researchers have found that exposure to electromagnetic fields actually promotes faster growth of cancer cells that are more resistant to anticancer drugs. Other studies indicate that exposure to electromagnetic fields inhibits human production of melatonin, a cancer-inhibiting hormone.

In addition, scientists are also trying to discover the actual mechanisms by which

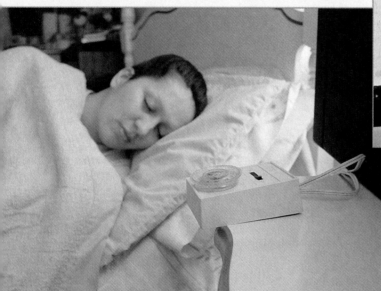

◀ ▲ **Low-level electromagnetic fields are created by common electrical devices such as hair dryers, electric blankets, and computers. Whether or not these fields are dangerous is yet to be proven.**

electromagnetic fields do their damage. They are focusing on the effect of electromagnetic fields on the flow of ions into and out of cells. Some experiments have shown that the fields may resonate (vibrate at the same frequency) with ions already present in the cell. Such vibration causes the valuable ions to pass through the membrane at an increased rate, thus leaving the cell too quickly and possibly damaging the cell membrane. Other research suggests that under certain conditions, the interaction of the Earth's magnetic field with unnatural electromagnetic fields may knock ions in the cell membrane out of place. This disrupts cell functioning, perhaps leading to illness.

The controversy over the safety or threat of electromagnetic radiation becomes increasingly serious as time goes on and society relies more and more on electric components that produce electromagnetic radiation. Until answers and resolutions to the problem are found, we shall continue to use and rely on electricity to assist and enhance our lives. And in one sense, we shall continue to be human guinea pigs in an unresolved scientific experiment.

THE COMPUTER THAT LIVES!

Ronda and her family lived in a space settlement on Pluto. One day, a strong radiation storm swept across the Purple Mountains of their planet. There had been many such storms on the planet in the year 2101. The module in which Ronda and her family lived had been directly in the path of the storm. Somehow, radioactive dust penetrated the sealed glass that served as windows in the module. As a result, Ronda was blind. Radiation had destroyed the nerves that carried the electrical signals from Ronda's eyes to her brain. Her brain could no longer interpret what her eyes were "seeing clearly."

Months after the storm, Ronda sat nervously in a plush armchair in the waiting room of Venus General Hospital. Today was the day the bandages would be removed from her eyes. Ronda was terrified that the operation to restore her sight might have been a failure. She did not want to rely on a seeing-eye robot for the rest of her life.

As the doctors removed the bandages, Ronda thought about the computer that had been implanted in her brain. No larger than a grain of rice, the computer was programmed to record all the images Ronda's eyes picked up and then translate them into messages her brain could understand. The computer was designed to work exactly like the eye nerves that had been destroyed.

The bandages fell from Ronda's eyes. She could see! The living computer inside her head had restored her sight.

TIME MARCHES ON

Today scientists believe that living computers will be a reality in the not-so-distant future. Living computers, like the one in Ronda's brain, require no outside power source and never need to be replaced. To understand how living computers may be possible, let's look briefly at how computers have evolved since the 1800s.

The first computers, made up of clunky gears and wheels, were turned by hand. They had only the simplest ability to answer questions based on the information stored inside them. By 1950, computers were run by electricity and operated with switches instead of gears. Information was stored when thousands of switches turned on and off in certain ways. While a lot of information could be fed into electrical memory banks, the computers of the 1950s were still very crude. In fact, the typical 1950s computer took up an entire room and could do less than many video-arcade games of the 1980s.

▲ Vacuum tubes like these made the first computers possible.

▲ The computers of the 1950s used transistors instead of vacuum tubes.

The computers of the 1980s contain thousands of switches and can be placed on a tabletop. The reason for the compactness of these computers is the silicon chip. Engineers can put hundreds of switches on a tiny piece of the chemical silicon. These tiny pieces of silicon, known as chips, are manufactured using laser beams and microscopes. It is these chips that record the information when, for example, you tell a computer your name.

But even with the silicon chip, modern computers cannot really think creatively or reason. And a computer has less sense than an ordinary garden snail. Experts feel that in order for computers to "graduate" to higher-level tasks, a whole new way must be developed of storing information in them. The key to developing a new system of information storage may lie in molecules of certain chemicals.

Why molecules? Scientists know that when some molecules are brought together, interesting changes take place. For example, electricity can jump from one molecule to another almost as if tiny switches were being shut on and off between them. Can we learn how to work these tiny switches? If so, then perhaps a whole new type of computer could be built!

This new computer might be able to hold more information in a single drop of liquid

▶ When nerves connecting the eye to the brain are destroyed, no electrical signals can be carried. A person cannot see. By implanting a computer the size of a grain of rice, the person's sight is restored. The computer is designed to work exactly like the eye nerves.

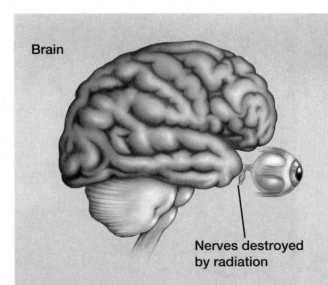

Brain

Nerves destroyed by radiation

▲ Today, thousands of transistors are packed onto a silicon chip.

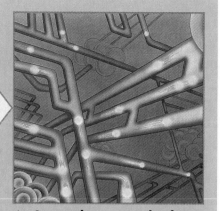

▲ A complex network of protein molecules, such as the one in this drawing, may be the electrical switches in computers of the future.

every minute of every day. Some scientists feel that bacteria can be "taught" to make special molecules. These molecules, when mixed together, could produce the flow of electricity needed to make a computer.

Of course, bacteria cannot be taught in the same sense that people can. Bacteria are not able to "learn." However, scientists can now control bacteria in many unusual ways. There are new techniques available that allow scientists to combine two different types of bacteria to produce a third, totally different, type. In the future, bacteria may produce chemicals that have never been seen before.

than today's computer could store in an entire roomful of chips. As you can imagine, the molecules in this computer of the future would have to be pretty special. And they would have to be produced in a new way.

LEAVE IT TO BACTERIA

One of the most popular current ideas concerning how these molecules could be produced is: Let bacteria do it for us! Bacteria are all around us. They constantly break down very complicated chemicals into molecules. Bacteria are at work in our bodies, in our food, and in our environment

In terms of a living computer, imagine that some bacteria have been taught to make special molecules. These bacteria could be grown in a special container and fed a particular substance to produce certain molecules. If the molecules could be told, or programmed, to do the right things, you would have a computer. And the computer would actually be alive because the bacteria would live, grow, and produce molecules inside their container.

Think about the living computer implanted in Ronda's brain, which allowed her to see again. Remember that bacteria need food to make molecules. Suppose that the computer in Ronda's brain was fed by her own blood, like all the other cells in her body. If this were the case, Ronda's computer would live as long as Ronda herself.

It may be many years before the living computer becomes a reality. Scientists must learn more about such things as how molecules react together and how they can be programmed. But many scientists await the day when they can look at a computer and say, "It's alive."

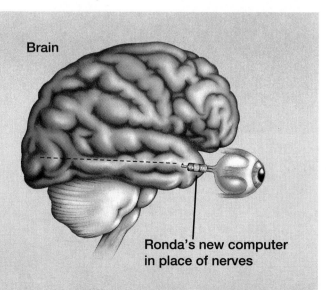

Brain

Ronda's new computer in place of nerves

For Further Reading

> If you have been intrigued by the concepts examined in this textbook, you may also be interested in the ways fellow thinkers—novelists, poets, essayists, as well as scientists—have imaginatively explored the same ideas.

Chapter 1: Electric Charges and Currents

Cosner, Sharon. *The Light Bulb: Inventions that Changed Our Lives.* New York: Walker.

Franklin, Benjamin. *Autobiography of Benjamin Franklin.* New York: Airmont.

Shelley, Mary. *Frankenstein.* London, England: Penguin.

Silverstein, Shel. *A Light in the Attic.* New York: Harper & Row.

Chapter 2: Magnetism

Averons, Pierre. *The Atom.* New York: Barron.

Clark, Electra. *Robert Peary: Boy of the North Pole.* New York: Macmillan.

Clarke, Arthur C. *The Wind from the Sun: Stories of the Space Age.* New York: New American Library.

Mason, Theodore K. *Two Against the Ice: Amundsen and Ellsworth.* Spring Valley, NY: Dodd, Mead.

Vogt, Gregory. *Electricity and Magnetism.* New York: Watts.

Chapter 3: Electromagnetism

Ellis, Ella T. *Riptide.* New York: Macmillan.

Evans, Arthur N. *The Automobile.* Minneapolis, MN: Lerner Publications.

Farr, Naunerle C. *Thomas Edison—Alexander Graham Bell.* West Haven, CT: Pendulum Press.

Mazer, Harry. *When the Phone Rang.* New York: Scholastic.

Snow, Dorothea J. *Samuel Morse: Inquisitive Boy.* New York: Macmillan.

Chapter 4: Electronics and Computers

Chetwin, Grace. *Out of the Dark World.* New York: Lothrop, Lee & Shepard Books.

Clarke, Arthur C. *2001: A Space Odyssey.* New York: New American Library.

Francis, Dorothy. *Computer Crime.* New York: Lodestar.

Trainer, David. *A Day in the Life of a TV News Reporter.* Mahwah, NJ: Troll.

Appendix A

THE METRIC SYSTEM

The metric system of measurement is used by scientists throughout the world. It is based on units of ten. Each unit is ten times larger or ten times smaller than the next unit. The most commonly used units of the metric system are given below. After you have finished reading about the metric system, try to put it to use. How tall are you in metrics? What is your mass? What is your normal body temperature in degrees Celsius?

Commonly Used Metric Units

Length   The distance from one point to another

meter (m)   A meter is slightly longer than a yard.
1 meter = 1000 millimeters (mm)
1 meter = 100 centimeters (cm)
1000 meters = 1 kilometer (km)

Volume   The amount of space an object takes up

liter (L)   A liter is slightly more than a quart.
1 liter = 1000 milliliters (mL)

Mass   The amount of matter in an object

gram (g)   A gram has a mass equal to about one paper clip.

1000 grams = 1 kilogram (kg)

Temperature   The measure of hotness or coldness

degrees   0°C = freezing point of water
Celsius (°C)   100°C = boiling point of water

Metric–English Equivalents

2.54 centimeters (cm) = 1 inch (in.)
1 meter (m) = 39.37 inches (in.)
1 kilometer (km) = 0.62 miles (mi)
1 liter (L) = 1.06 quarts (qt)
250 milliliters (mL) = 1 cup (c)
1 kilogram (kg) = 2.2 pounds (lb)
28.3 grams (g) = 1 ounce (oz)
$°C = 5/9 \times (°F - 32)$

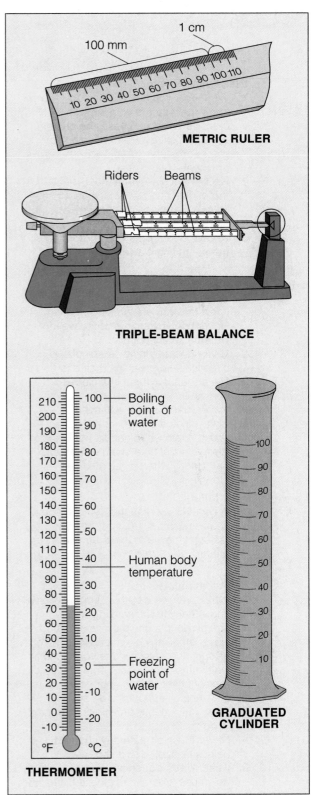

METRIC RULER

TRIPLE-BEAM BALANCE

THERMOMETER

GRADUATED CYLINDER

Appendix B

Glassware Safety

1. Whenever you see this symbol, you will know that you are working with glassware that can easily be broken. Take particular care to handle such glassware safely. And never use broken or chipped glassware.
2. Never heat glassware that is not thoroughly dry. Never pick up any glassware unless you are sure it is not hot. If it is hot, use heat-resistant gloves.
3. Always clean glassware thoroughly before putting it away.

Fire Safety

1. Whenever you see this symbol, you will know that you are working with fire. Never use any source of fire without wearing safety goggles.
2. Never heat anything—particularly chemicals—unless instructed to do so.
3. Never heat anything in a closed container.
4. Never reach across a flame.
5. Always use a clamp, tongs, or heat-resistant gloves to handle hot objects.
6. Always maintain a clean work area, particularly when using a flame.

Heat Safety

Whenever you see this symbol, you will know that you should put on heat-resistant gloves to avoid burning your hands.

Chemical Safety

1. Whenever you see this symbol, you will know that you are working with chemicals that could be hazardous.
2. Never smell any chemical directly from its container. Always use your hand to waft some of the odors from the top of the container toward your nose—and only when instructed to do so.
3. Never mix chemicals unless instructed to do so.
4. Never touch or taste any chemical unless instructed to do so.
5. Keep all lids closed when chemicals are not in use. Dispose of all chemicals as instructed by your teacher.

6. Immediately rinse with water any chemicals, particularly acids, that get on your skin and clothes. Then notify your teacher.

Eye and Face Safety

1. Whenever you see this symbol, you will know that you are performing an experiment in which you must take precautions to protect your eyes and face by wearing safety goggles.
2. When you are heating a test tube or bottle, always point it away from you and others. Chemicals can splash or boil out of a heated test tube.

Sharp Instrument Safety

1. Whenever you see this symbol, you will know that you are working with a sharp instrument.
2. Always use single-edged razors; double-edged razors are too dangerous.
3. Handle any sharp instrument with extreme care. Never cut any material toward you; always cut away from you.
4. Immediately notify your teacher if your skin is cut.

Electrical Safety

1. Whenever you see this symbol, you will know that you are using electricity in the laboratory.
2. Never use long extension cords to plug in any electrical device. Do not plug too many appliances into one socket or you may overload the socket and cause a fire.
3. Never touch an electrical appliance or outlet with wet hands.

Animal Safety

1. Whenever you see this symbol, you will know that you are working with live animals.
2. Do not cause pain, discomfort, or injury to an animal.
3. Follow your teacher's directions when handling animals. Wash your hands thoroughly after handling animals or their cages.

One of the first things a scientist learns is that working in the laboratory can be an exciting experience. But the laboratory can also be quite dangerous if proper safety rules are not followed at all times. To prepare yourself for a safe year in the laboratory, read over the following safety rules. Then read them a second time. Make sure you understand each rule. If you do not, ask your teacher to explain any rules you are unsure of.

Dress Code

1. Many materials in the laboratory can cause eye injury. To protect yourself from possible injury, wear safety goggles whenever you are working with chemicals, burners, or any substance that might get into your eyes. Never wear contact lenses in the laboratory.

2. Wear a laboratory apron or coat whenever you are working with chemicals or heated substances.

3. Tie back long hair to keep it away from any chemicals, burners and candles, or other laboratory equipment.

4. Remove or tie back any article of clothing or jewelry that can hang down and touch chemicals and flames.

General Safety Rules

5. Read all directions for an experiment several times. Follow the directions exactly as they are written. If you are in doubt about any part of the experiment, ask your teacher for assistance.

6. Never perform activities that are not authorized by your teacher. Obtain permission before "experimenting" on your own.

7. Never handle any equipment unless you have specific permission.

8. Take extreme care not to spill any material in the laboratory. If a spill occurs, immediately ask your teacher about the proper cleanup procedure. Never simply pour chemicals or other substances into the sink or trash container.

9. Never eat in the laboratory.

10. Wash your hands before and after each experiment.

First Aid

11. Immediately report all accidents, no matter how minor, to your teacher.

12. Learn what to do in case of specific accidents, such as getting acid in your eyes or on your skin. (Rinse acids from your body with lots of water.)

13. Become aware of the location of the first-aid kit. But your teacher should administer any required first aid due to injury. Or your teacher may send you to the school nurse or call a physician.

14. Know where and how to report an accident or fire. Find out the location of the fire extinguisher, phone, and fire alarm. Keep a list of important phone numbers—such as the fire department and the school nurse—near the phone. Immediately report any fires to your teacher.

Heating and Fire Safety

15. Again, never use a heat source, such as a candle or burner, without wearing safety goggles.

16. Never heat a chemical you are not instructed to heat. A chemical that is harmless when cool may be dangerous when heated.

17. Maintain a clean work area and keep all materials away from flames.

18. Never reach across a flame.

19. Make sure you know how to light a Bunsen burner. (Your teacher will demonstrate the proper procedure for lighting a burner.) If the flame leaps out of a burner toward you, immediately turn off the gas. Do not touch the burner. It may be hot. And never leave a lighted burner unattended!

20. When heating a test tube or bottle, always point it away from you and others. Chemicals can splash or boil out of a heated test tube.

21. Never heat a liquid in a closed container. The expanding gases produced may blow the container apart, injuring you or others.

22. Before picking up a container that has been heated, first hold the back of your hand near it. If you can feel the heat on the back of your hand, the container may be too hot to handle. Use a clamp or tongs when handling hot containers.

Using Chemicals Safely

23. Never mix chemicals for the "fun of it." You might produce a dangerous, possibly explosive substance.

24. Never touch, taste, or smell a chemical unless you are instructed by your teacher to do so. Many chemicals are poisonous. If you are instructed to note the fumes in an experiment, gently wave your hand over the opening of a container and direct the fumes toward your nose. Do not inhale the fumes directly from the container.

25. Use only those chemicals needed in the activity. Keep all lids closed when a chemical is not being used. Notify your teacher whenever chemicals are spilled.

26. Dispose of all chemicals as instructed by your teacher. To avoid contamination, never return chemicals to their original containers.

27. Be extra careful when working with acids or bases. Pour such chemicals over the sink, not over your workbench.

28. When diluting an acid, pour the acid into water. Never pour water into an acid.

29. Immediately rinse with water any acids that get on your skin or clothing. Then notify your teacher of any acid spill.

Using Glassware Safely

30. Never force glass tubing into a rubber stopper. A turning motion and lubricant will be helpful when inserting glass tubing into rubber stoppers or rubber tubing. Your teacher will demonstrate the proper way to insert glass tubing.

31. Never heat glassware that is not thoroughly dry. Use a wire screen to protect glassware from any flame.

32. Keep in mind that hot glassware will not appear hot. Never pick up glassware without first checking to see if it is hot. See #22.

33. If you are instructed to cut glass tubing, fire-polish the ends immediately to remove sharp edges.

34. Never use broken or chipped glassware. If glassware breaks, notify your teacher and dispose of the glassware in the proper trash container.

35. Never eat or drink from laboratory glassware. Thoroughly clean glassware before putting it away.

Using Sharp Instruments

36. Handle scalpels or razor blades with extreme care. Never cut material toward you; cut away from you.

37. Immediately notify your teacher if you cut your skin when working in the laboratory.

Animal Safety

38. No experiments that will cause pain, discomfort, or harm to mammals, birds, reptiles, fishes, and amphibians should be done in the classroom or at home.

39. Animals should be handled only if necessary. If an animal is excited or frightened, pregnant, feeding, or with its young, special handling is required.

40. Your teacher will instruct you as to how to handle each animal species that may be brought into the classroom.

41. Clean your hands thoroughly after handling animals or the cage containing animals.

End-of-Experiment Rules

42. After an experiment has been completed, clean up your work area and return all equipment to its proper place.

43. Wash your hands after every experiment.

44. Turn off all burners before leaving the laboratory. Check that the gas line leading to the burner is off as well.

Glossary

alternating current: current in which the electrons reverse their direction regularly

amplifier: device that increases the strength of an electric signal

atom: smallest part of an element that has all the properties of that element

aurora: glowing region of air caused by solar particles that break through the Earth's magnetic field

battery: device that produces electricity by converting chemical energy into electrical energy; made up of electrochemical cells

binary system: number system consisting of only two numbers, 0 and 1, that is used by computers

bit: single electronic switch, or piece of information

byte: string of bits; usually 8 bits make up a byte

cathode-ray tube: type of vacuum tube that uses electrons to produce an image on a screen

central processing unit: part of a computer that controls the operation of all the other components of the computer

charge: physical property of matter that can give rise to an electric force of attraction or repulsion

chip: thin piece of silicon containing an integrated circuit

circuit: complete path through which electricity can flow

circuit breaker: reusable device that protects a circuit from becoming overloaded

conduction: method of charging an object by allowing electrons to flow through one object to another object

conductor: material which permits electrons to flow freely

current: flow of charge

diode: vacuum tube or semiconductor that acts as a rectifier

direct current: current consisting of electrons that flow constantly in one direction

disk drive: part of a computer that can act as an input device by reading information off a disk and entering it into the computer or as an output device removing information from a computer and storing it on a disk

doping: process of adding impurities to semiconducting materials

electric discharge: loss of static electricity as electric charges move off an object

electric field: area over which an electric charge exerts a force

electric motor: device that uses an electromagnet to convert electrical energy to mechanical energy that is used to do work

electromagnet: solenoid with a magnetic material such as iron inside its coils

electromagnetic induction: process by which a current is produced by a changing magnetic field

electromagnetic wave: wave made up of a combination of a changing electric field and a changing magnetic field

electromagnetism: relationship between electricity and magnetism

electron: subatomic particle with a negative charge found in an area outside the nucleus of an atom

electronics: study of the release, behavior, and control of electrons as it relates to use in practical devices

electroscope: instrument used to detect charge

force: push or pull on an object

friction: force that opposes motion that is exerted when two objects are rubbed together in some way

fuse: thin strip of metal used for safety because when the current flowing through it becomes too high, it melts and breaks the flow of electricity

galvanometer: device that uses an electromagnet to detect small amounts of current

generator: device that uses electromagnets to convert mechanical energy to electrical energy

hardware: physical parts of a computer

induced current: current produced in a wire exposed to a changing magnetic field

induction: method of charging an object by rearranging its electric charges into groups of positive charge and negative charge

input device: device through which data is fed into a computer

insulator: material made up of atoms with tightly bound electrons that are not able to flow freely

integrated circuit: circuit consisting of many diodes and transistors all placed on a thin piece of silicon, known as a chip

magnetic domain: region of a material in which the magnetic fields of individual atoms are aligned

magnetic field: area over which the magnetic force is exerted

magnetism: force of attraction or repulsion of a magnetic material due to the arrangement of its atoms

magnetosphere: region in which the magnetic field of the Earth is found

main memory: part of a computer that contains data and operating instructions that are processed by the central processing unit

modem: device that changes electronic signals from a computer into messages that can be carried over telephone lines

neutron: subatomic particle with no charge located in the nucleus of an atom

Ohm's law: expression that relates current, voltage, and resistance: $V = I \times R$

output device: part of a computer through which information is removed

parallel circuit: circuit in which different parts are on separate branches; if one part does not operate properly, current can still flow through the others

photocell: device that uses electrons emitted from a metal during the photoelectric effect to produce current

pole: regions of a magnet where the magnetic effects are the strongest

potential difference: difference in charge as created by opposite posts of a battery

power: rate at which work is done or energy is used

proton: subatomic particle located in the nucleus of an atom with a positive charge

rectifier: device that converts alternating current to direct current; accomplished by a vacuum tube or semiconductor called a diode

resistance: opposition to the flow of electric charge

semiconductor: material that is able to conduct electric currents better than insulators but not as well as true conductors

series circuit: circuit in which all parts are connected one after another; if one part fails to operate properly, the current cannot flow

software: set of instructions, or program, a computer follows

solenoid: long coil of wire that acts like a magnet when current flows through it

solid-state device: device, consisting of semiconductors, that has come out of the study of the structure of solid materials

static electricity: movement of charges from one object to another without further movement

superconductor: material in which resistance is essentially zero at certain low temperatures

thermocouple: device that produces electrical energy from heat energy

transformer: device that increases or decreases the voltage of alternating current

transistor: device consisting of three layers of semiconductors used to amplify an electric signal

triode: type of vacuum tube used for amplification that consists of a wire grid as well as its electrodes

vacuum tube: glass tube, in which almost all gases are removed, which contains electrodes that produce a one-way flow of electrons

voltage: potential difference; energy carried by charges that make up a current

Index

Credits

Cover Background: Ken Karp
Photo Research: Omni-Photo Communications, Inc.
Contributing Artists: Illustrations: Warren Budd Assoc. Ltd., Gerry Schrenk, Martinu Schneegass. Charts and graphs: Function Thru Form
Photographs: 4 top: Phil Degginger; bottom left: Dennis Purse/Photo Researchers, Inc.; bottom right: Michael Philip Manheim; **5** top: Dr. Jeremy Burgess/Science Photo Library/Photo Researchers, Inc.; bottom: NASA; **6** top: Lefever/Grushow/Grant Heilman Photography; center: Index Stock Photography, Inc.; bottom: Rex Joseph; **8** top: Kobal Collection/Superstock; bottom: Jerry Mason/Science Photo Library/Photo Researchers, Inc.; **9** Hank Morgan/Science Source/Photo Researchers, Inc.; **12** Robert Western/Tony Stone Worldwide/Chicago Ltd.; **15** Fundamental Photographs; **16** Michael Philip Manheim; **18** Phil Jude/Science Photo Library/Photo Researchers, Inc.; **19** top: North Wind Picture Archives; bottom: Tony Stone Worldwide/Chicago Ltd.; **20** left: G. V. Faint/Image Bank; right: A. d'Arazien/Image Bank; **23** Paul Shambroom/Photo Researchers, Inc.; **25** left: J. Alex Langley/DPI; right: Henry Grossman/DPI; **26** Ron Scott/Tony Stone Worldwide/Chicago Ltd.; **27** Bob Evans/Peter Arnold, Inc.; **28** Gary Gladstone/Image Bank; **29** Phil Degginger; **32** Brian Parker/Tom Stack & Associates; **34** Paul Silverman/Fundamental Photographs; **37** Ken Karp; **43** R. J. Erwin/Photo Researchers, Inc.; **44 and 45** Fundamental Photographs; **46, 47** left and right, **48, 49** top and bottom, and **51** Richard Megna/Fundamental Photographs; **52** GE Corporate Research and Development; **53** Dr. E. R. Degginger; **54** top: Granger Collection; bottom: Francois Gohier/Photo Researchers, Inc.; **55** Science Photo Library/Photo Researchers, Inc.; **57** Jack Finch/Science Photo Library/Photo Researchers, Inc.; **58** Max-Planck-Institut Fur Radioastronomie/Science Photo Library/Photo Researchers, Inc.; **64 and 65** Kaku Kurita/Gamma-Liaison, Inc.; **68** left: Dick Durrance II/Woodfin Camp & Associates; right: Ken Karp; **69** Don Klumpp/Image Bank; **76** left: Brian Parker/Tom Stack & Associates; right: D.O.E./Science Source/Photo Researchers, Inc.; **77** Richard Megna/Fundamental Photographs; **79** Dr. E. R. Degginger; **81** top and bottom: Dr. Jeremy Burgess/Science Photo Library/Science Photo Library/Photo Researchers, Inc.; **85** U. S. Department of the Interior, National Park Service, Edison National Historic Site/Omni-Photo Communications, Inc.; **86 and 87** Joel Gordon; **88** top left: Mitchell Bleier/Peter Arnold, Inc.; bottom left: Dennis Purse/Photo Researchers, Inc.; bottom right: Walter Bibikow/Image Bank; **90** Ken Karp; **91** Robert Matheu/Retna Limited; **92 and 93** Ken Karp; **94** left: Chuck O'Rear/Woodfin Camp & Associates; center: Alfred Pasieka/Peter Arnold, Inc.; right: Joel Gordon; **96** top left: Biophoto Associates/Photo Researchers, Inc.; top right: Petit Format/Guigoz/Steiner/Science Source/Photo Researchers, Inc.; center: Morton Beebe/Image Bank; bottom: Stephenie S. Ferguson; **98** Culver Pictures, Inc.; **99** Dan McCoy/Rainbow; **102** left: IBM; right: Granger Collection; **103** top: Srulik Haramaty/Phototake; bottom left: NASA; bottom right: Gregory Sams/Science Photo Library/Photo Researchers, Inc.; **104** Ken Karp/Omni-Photo Communications, Inc.; **105** Eric Kroll/Omni-Photo Communications, Inc.; **111** Phil Degginger; **112** left and right: IBM Research; **113** Chris Rogers/Stock Market; **114** Peter Poulides/Tony Stone Worldwide/Chicago Ltd.; **115** left: Mike Borum/Image Bank; right: Barry Lewis/Tony Stone Worldwide/Chicago Ltd.; right inset: Scott Camazine/Photo Researchers, Inc.; **116** top left: Hank Morgan/Photo Researchers, Inc.; right: Ken Lax/Photo Researchers, Inc.; bottom left: Tony Freeman/Photoedit; **118** left: Dan McCoy/Rainbow; right: Dick Luria/Science Source/Photo Researchers, Inc.; **119** Roger Du Buisson/Stock Market; **120** Kaku Kurita/Gamma-Liaison, Inc.